worthy pursuit close to my heart, and I was fortunate to have
There are many grave issues that stare menacingly in the face of
deserve immediate attention. The issue of growing waste ranks at the
negative implications for environmental preservation and conservation
nomic prosperity, and for continued soundness of human health and
mind, the issue of growing waste needs to be tackled with fierce urgency,
fford the luxury of waiting one more minute to take this dillemma head
need to dedicate every iota of our resources and our beings to restrain
it. Kanczuzeweski's book provides a pragmatic and effective roadmap to
he gravity of the issue and to mount an effective attack."

**Ashok Kumar, Grand Valley State University emeritus professor
of management, and founder-chairman of Global Foundation
for Advancement of Environment and Human Wellness**

ntapped potential that could come from landfill mitigation is immense. Each
l stream contains value, as long as there is enough volume or a system in place to
it. Recycling and scrap industries have grown to be more efficient, but there is still
work to be done. We need more people to get into the business of reinventing waste,
use it is a valuable resource."

Kari Bliss, customer experience and sustainability at PADNOS

"We can no longer perpetu
away by landfilling them. We
we're throwing away, no longer
The reality is our current approac.
create externalities that are carried
some great options for communities

Darwin J. Baas, directo

"Tyler's passion for sustainability is only m.
strategies to reshape our view of waste and re
treatise addresses the critical changes we mus
importantly, our children's future, is healthier, cle
new generation of leaders are tackling growing threa
and Tyler is helping to lead this charge."

Chuck B

"In *Reinvent Your Waste*, Tyler turns his passion for sust
introduces us to the 4R Earth approach—respect, recover,
provides a rich compendium of resources, protocols, and practic
begin putting into action immediately. He offers us hope and a
which we can all join in the movement to create a brighter tomorro

Mike Keen, IU South Bend cha.
emeritus of sustainability studi

"This project was a
participated in it.
humanity today an
top. It has massive
of the planet, ec
wellness. To my
now. We can't
on; indeed, w
and vanquish
understand

"The u
materi
collec
much
beca

"This project was a worthy pursuit close to my heart, and I was fortunate to have participated in it. There are many grave issues that stare menacingly in the face of humanity today and deserve immediate attention. The issue of growing waste ranks at the top. It has massive negative implications for environmental preservation and conservation of the planet, economic prosperity, and for continued soundness of human health and wellness. To my mind, the issue of growing waste needs to be tackled with fierce urgency, now. We can't afford the luxury of waiting one more minute to take this dillemma head on; indeed, we need to dedicate every iota of our resources and our beings to restrain and vanquish it. Kanczuzeweski's book provides a pragmatic and effective roadmap to understand the gravity of the issue and to mount an effective attack."

Ashok Kumar, Grand Valley State University emeritus professor of management, and founder-chairman of Global Foundation for Advancement of Environment and Human Wellness

"The untapped potential that could come from landfill mitigation is immense. Each material stream contains value, as long as there is enough volume or a system in place to collect it. Recycling and scrap industries have grown to be more efficient, but there is still much work to be done. We need more people to get into the business of reinventing waste, because it is a valuable resource."

Kari Bliss, customer experience and sustainability at PADNOS

Remarks

"We can no longer perpetuate the idea that it's simply OK to throw all these resources away by landfilling them. We all individually make decisions daily that whatever it is that we're throwing away, no longer has value. As a community, we need to decide differently. The reality is our current approach and decisions, whether as a company or as individuals, create externalities that are carried by all of us. We all have a choice, and this book has some great options for communities to incorporate and embrace."

Darwin J. Baas, director at Kent County Department of Public Works

"Tyler's passion for sustainability is only matched by his curiosity and interest in societal strategies to reshape our view of waste and reverse factors leading to planet decline. His treatise addresses the critical changes we must make to ensure our future, and more importantly, our children's future, is healthier, cleaner, and sustainable. I'm excited that a new generation of leaders are tackling growing threats with a real blueprint and approach, and Tyler is helping to lead this charge."

Chuck Bower, Hawthorne Services, LLC

"In *Reinvent Your Waste*, Tyler turns his passion for sustainability to zero waste. He introduces us to the 4R Earth approach—respect, recover, reinvent, and restore—and provides a rich compendium of resources, protocols, and practical steps that any of us can begin putting into action immediately. He offers us hope and a proactive path through which we can all join in the movement to create a brighter tomorrow."

**Mike Keen, IU South Bend chancellor's professor
emeritus of sustainability studies and sociology**

Reinvent
Your
Waste

The 4-Stepped Plan and Call-to-Action Guide for Stewards to Reinvent and Revalue Waste

Tyler Kanczuzewski

ARCHWAY
PUBLISHING

This book is a work of non-fiction. Unless otherwise noted, the author and the publisher make no explicit guarantees as to the accuracy of the information contained in this book and in some cases, names of people and places have been altered to protect their privacy.

Archway Publishing books may be ordered through booksellers or by contacting:

Archway Publishing
1663 Liberty Drive
Bloomington, IN 47403
www.archwaypublishing.com
844-669-3957

Because of the dynamic nature of the Internet, any web addresses or links contained in this book may have changed since publication and may no longer be valid. The views expressed in this work are solely those of the author and do not necessarily reflect the views of the publisher, and the publisher hereby disclaims any responsibility for them.

ISBN: 978-1-6657-4344-0 (sc)
ISBN: 978-1-6657-4345-7 (e)

Library of Congress Control Number: 2023908063

Print information available on the last page.

Archway Publishing rev. date: 08/04/2023

Contents

A Waste Evolution Story

The Four-Step Plan
and Call-to-Action Guide
for Stewards to Reinvent and Revalue Waste

Respect, Recover, Reinvent, and Restore
4R Earth and for Michigan

This project is dedicated to my parents,
Tom and Lyrin Kanczuzewski,
who put me on this planet and path to better understand waste
and try to make the world a better place.

Key Acknowledgments and Recognitions

Thanks to everyone who helped on this journey to better understand waste, but specifically the municipal solid waste and general trash that is polluting our Earth. We were able to dive into some of the key data points and statistics around the waste dilemma and discuss potential solutions for reinventing waste. There is still much work to be done, and it will take a collective force to turn things around for the planet.

This book is also dedicated to the following individuals, to any steward, and to all earthlings:

Ashok Kumar, PhD—Grand Valley State University, project advisor.

Christina Miller—Michigan Department of Environmental Quality. She provided a phone and email interview about her work and experience in Michigan.

Chuck Bower—Hawthorne Services, LLC. He is a personal business coach and project supporter.

Darwin Baas—Kent County, Michigan Department of Public Works. He provided phone and email interviews and communication about waste issues and potential solutions, as well as the proposed business park in Michigan for better recycling.

Jaideep Motwani, PhD—Grand Valley State University, research supporter.

Kari Bliss—PADNOS. She provided a phone interview on recycling and scrap industry in Michigan and beyond.

Katie Venechuk—Michigan Department of Environmental Quality, for her phone interview about her work and experience in Michigan.

Lauren Westerman—Kent County Michigan Department of Public Works, for providing a tour and for information gathering at a Grand Rapids, Michigan, recycling facility.

Lee Saberson—Richard G. Lugar Center for Renewable Energy, Indiana University, Purdue University, Indianapolis. Information gathering and research supporter.

Mike Keen—Thrive Michiana. He is a sustainability consultant and research supporter.

Norman Christopher—Grand Valley State University and Sustainable Business Practices, LLC. He was a key project advisor.

Paul Smith—Kent County Michigan Department of Public Works, for providing a tour of a Covanta Waste-to-Energy Plant and conversation about waste-to-energy.

T. J. Kanczuzewski—Inovateus Solar, LLC, Mamoni Valley Preserve. He conducted phone and email interviews on sustainability initiatives such as renewable energy and forest conservation.

Tyler Ganus—Southeast Berrien County Landfill, for providing a tour of a landfill and education on landfills.

Thanks to everyone for contributing, and please enjoy the following research.

A Briefing for Those in a Hurry

Waste is not being used as a valuable resource. The growth in total waste is an indication that we have not taken the opportunity to optimize waste in Michigan, in the United States, and across the globe. Municipal solid waste (MSW)—and particularly product, material, and packaging waste—are quickly accumulating. Waste, or "trash," is growing at unprecedented rates, which calls for newer and bigger landfills. Not only do Michigan and the United States lack space for these enlarged or new landfills, society can't afford the environmental harm, economic loss, and deterioration of human health. According to the Environmental Protection Agency (EPA) in 2018, the United States generated 292.4 million tons of MSW, which is more waste than ever recorded. More than 50 percent (146.2 million tons) of that waste went to landfills across the country, and approximately 32 percent was recycled and composted.

Is it cheaper to send waste to a landfill or to recycle and reuse it? Studies show that the United States and most citizens of Earth are living out of balance with the natural world and consuming resources faster than they can regenerate them. According to an ecological footprint calculator by Global Footprint Network, the United States alone needs about five Earths, and the entire world population needs roughly 1.7 Earths to keep up with current human consumption of resources and biodiversity. At the same time, global waste, pollution, and the population are growing exponentially. Because of this dichotomy, it makes sense to value waste much differently. Waste can be used and reinvented as a valuable resource to help save the planet as well as to create new business platforms and economies focused on stewardship, sustainability practices, and proper waste (resource) management.

Numerous movements, initiatives, and policies have been implemented to spread awareness and to help people think differently about waste conservation and recovery. One great initiative was the 3R mindset adopted in the 1970s: reduce, reuse, recycle. The 3R mindset worked tremendously well and is still used today. The '70s in general were a pivotal decade for creating much-needed sustainable mindset shifts to fight the growing epidemic of waste and pollution. Now, forty-plus years later, sustainability efforts have grown, and waste management is gradually improving. However, are these sustainable initiatives most efficient and timely enough, and can sustainable waste management radically improve? The short-and-sweet answer and belief is yes. We can greatly reduce the amount of waste going to landfills, litter that is saturating our lands, and trash that is polluting our precious waters. By adopting even newer mindsets, anything is possible, and hopefully humans can reverse these negative impacts.

The time is ripe for a mindset shift toward more innovative waste management. One idea is a four-step plan: respect, recover, reinvent, and restore (4R Earth). The four steps for Earth will first help establish a new approach to respecting the natural world and recover waste and resources to the best of our abilities. Next, reinventing waste is introduced to promote using waste as a valuable resource. The last step is to restore, the idea being to let Earth heal and to encourage earthlings to live more in balance with the planet, assuring that it provides for all future generations. Over the last decade, great awareness has been spread throughout the planet, teaching people about the growing accumulation of waste and pollution year over year. Waste from single-use disposables, packaging, and plastics are known to be especially high. New initiatives, movements, and policies are being implemented across the country throughout various thought-leading organizations. The four-step plan will help add to a foundation of historical sustainability and give birth to a new mindset shift that will further reinvent waste as a valuable resource. Waste is potentially the most undervalued resource of our time, and now is the time to reinvent it and value it like responsible stewards. The following information can also help any individual with sustainability strategy.

Please note: This research does not include industrial or construction and demolition (C&D) debris.

Foreword

Norman Christopher

Waste is an ongoing conversation, whether it involves a business, enterprises, or even a shopper. Waste has always been viewed as unavoidable, or as an outcome from the old "take-make-waste" extractive model. As either producers or consumers, we take in raw materials, make products, and generate waste. But there is a new, innovative approach in which to look at waste, and that is as regenerative raw material that can be innovatively developed and used to create value in the marketplace. *Reinvent Your Waste* addresses the changes needed to tackle the subject of waste via a different thought process. Whether you are a business, a municipality, a nonprofit, a service provider, or a consumer, there is something in this book from which you can learn.

The genesis of this book started with a journey that Tyler Kanczuzewski took while completing his MBA with an emphasis on sustainability at Grand Valley State University. He was an enthusiastic student who wanted to learn more about sustainability and about how he could walk the talk in his own life. Waste was a troublesome issue for Tyler, and he wanted to do additional independent research on the subject. Through his hard work and dedicated efforts, he has completed this book, *Reinvent Your Waste*. The purpose of this text is for us all to gain a greater understanding of the waste dilemma we are facing and the many issues surrounding this complex topic, and to appreciate some of the creative strategies being developed today to deal with waste, with the overall goal of leaving this world a better place in the future.

Tyler takes a refreshing and holistic methodology to address comprehensive waste issues. His work looks at waste through a new lens. Everyone creates waste, so everyone needs to take ownership for waste generation. With that ownership comes the responsibility of seeking innovative ways to make a difference in our own lives and to have influence where we work to reduce our collective waste generation.

The framework for this is rooted in stewardship and sustainable development best practices. This framework also embraces circularity and provides new insight on how to break down waste into four interconnective phases or steps:

1. Respecting our Earth
2. Recovering our resources
3. Reinventing our waste
4. Restoring our natural ecosystems

The first phase provides a thorough root-cause analysis of waste problems, including our ecological footprints and our collective consumption of scarce resources. The second phase deals with waste from compliance to regulation. Aggregate consumption of waste is provided by category. We will explore various current methodologies to deal with waste and a history and evolution of more sanitary waste management. The third phase takes a deeper dive into the protocols of Michigan's circular economy. The way we look at waste is transformed by viewing waste as a valuable raw material that can be innovatively and creatively used. Many resources are provided for specific sectors and organizations that are reinventing waste, as well as reference materials and books that raise the bar to generating zero waste. Tyler provides tips and ideas for how to change our lifestyles and mindsets, as well as thoughts about how to take corrective steps. Phase four focuses on restoring our planet and natural ecosystem as well as improving resiliency, highlighting the importance of taking crucial climate action steps and continuing the development of new technologies for transformational change.

This book acknowledges that hope and help is on the way. Best practices in waste management and sustainable development are available, and they call us to action to take corrective steps. I know you will find an action step that you can take to make this world a better place. Enjoy the book!

Introductions

There is a lot of talk about waste—especially food and plastic waste—that is being sent to our landfills and polluting our water and land. The big threat or unknown is, what is the growing waste doing to our natural ecosystems and to Earth's ancient systems? Recent scientific discoveries are showing that waste is negatively impacting the natural world and is likely disturbing life on Earth. We will discuss whether the negative impacts can be reversed and the progress that has been made or that is in the works to balance living with nature. We will also discuss whether waste—particularly material solid waste and products, material and packaging waste—is being properly valued and used. Every year, millions of tons of waste go to landfills across the country and are considered safely buried to decompose. To be exact, we sent 146.2 million tons of MSW and 144 million tons of C&D debris to our landfills in 2018. Are there more innovative ways to recover waste and create greater value, economically speaking?

This study hits on many topics, but all are aligned to handle waste in a more sustainable and steward-like way. But what does *sustainable* even mean? The roots of sustainability will be discussed first to lay the foundation for what sustainable waste management could be or strive to be. Then we will get into the heart of the study, which presents the idea to create a mindset shift regarding waste. The new idea, or mindset, is a stepped plan for how to live in balance with the natural world, to live more like a steward, and to think about waste as a valuable resource.

The four steps are respect, recover, reinvent, and restore (4R Earth). Step 1, respect, is associated with the natural world and discusses how our recent revolutions have helped speed things up in terms of human consumption and the development of various resources. Step 1 also interprets whether current demand of resources meets the Earth's natural supply. Step 2, recover, dives into what waste is and how management has evolved. Recovery of waste has been growing more sustainable in many ways; however, there are tremendous opportunities to achieve improved sustainability. Step 3, reinvent, first addresses the dilemma that the United States—and most of the world—is living out of balance with the natural world, while burying and not valuing waste. Step 3 then explores ideas and methods for any person, steward, or sustainability connoisseur to implement ideas to reinvent waste and help save the Earth, while using waste with higher value. Last, step 4, restore, closes with the idea that if proper steps are taken, we can regain a sense of balance with Earth and restore it to a sustainable baseline. We only have one planet, so we must try our best to live in proper balance with it. Earth is resilient, and if we respect it, we can restore it so that it provides life for all future generations.

After each step, there are tips and recommendations for stewards. A steward is someone who strives to do what is right and implements best practices into his or her life to be a sustainable champion and leader for his or her family, business, organization, or local community. This study raises awareness, provides ideas, and creates new general sustainability mindsets or concepts to embrace and live by for those individuals who want to be better stewards. This study is a steward's guide for reinventing waste. If we don't reinvent it, we will continue to negatively impact the natural world, which could be extremely dangerous and is a road humanity should not dare to venture down.

Reminder: This research does not include industrial or C&D debris. For example, in 2018, 600 million tons of C&D debris was recorded, and 144 million tons of that went to landfills, along with the 146.2 million tons of MSW. Many of the same issues and steward recommendations can be applied to industrial and C&D debris.

The Roots of Sustainability

Critical Dates: Before 1500, 1500, 1713, 1798, 1968, 1970s, 1987, 1994, 2000s, the future?

The foundation of sustainability has been in our human vocabulary for a long time. It is important to understand what sustainability means and why it is important. Therefore, it is important to explore the timeline and evolutionary history of the idea and philosophy of sustainability.

The word *sustain* has supposedly been in existence for over five hundred years. The word comes from Middle English times and has a Latin and French origin, meaning "to hold." Later interpretations of the word became known as "to make something continue or provide enough of what somebody needs in order to live" (Oxford Learner's Dictionary, n.d.). Then, in 1713, the German word *Nachhaltigkeith*, meaning "sustained yield," appeared in a handbook of forestry that was published that year. In that handbook, the term meant to never harvest more than the forest can regenerate (The World Energy Foundation 2015). Moving forward to 1798, Thomas Malthus wrote an essay called "Principal of Population," in which he brought about the idea of population growth outrunning available production and resources (Encyclopedia Britannica 1998). The philosophy Malthus brought about over two hundred years ago is now rooted in our modern-day concept of sustainability.

Native Americans have also been known for their unique philosophies of life and ways to live in peace and harmony with Earth. Some tribes are known to think seven generations ahead. There is an ancient Indian proverb that states: "Treat the Earth well. It was not given to you by your parents. It was loaned to you by your children." (Christopher 2012, 11). Most Native Americans were living and practicing sustainably before the word *sustainability* was being used. Native Americans and other indigenous tribes arguably planted our sustainability roots thousands of years ago.

Fast-forwarding to 1969, the US government enacted the National Environmental Policy Act (NEPA). The purpose of this act was to declare a national policy that would encourage productive and enjoyable harmony between humanity and its environment; to promote efforts that will prevent or eliminate damage to the environment and biosphere, and to stimulate the health and welfare of humankind; to enrich the understanding of the ecological systems and natural resources important to the nation; and to establish a Council on Environmental Quality (U.S. Department of Energy n.d.).

Not too long after NEPA was enacted, on April 22, 1970, US Senator of Wisconsin at that time, Gaylord Nelson, along with a large committee and task force, created a national movement called Earth Day. On that day, according to Earth Day Network, twenty million Americans took to the streets, parks, and auditoriums to demonstrate for a healthy, sustainable environment in massive coast-to-coast rallies. Thousands of colleges and universities organized protests against the deterioration of the environment. Groups that had been fighting against oil spills, polluting factories and power plants, raw sewage, toxic dumps, pesticides, freeways, the loss of wilderness, and the extinction of wildlife suddenly realized they shared common values (Earth Day Network 2018). The Earth Day Network manages the Earth Day movement and organizes annual events that are growing awareness. It is believed that the booming success and awareness that Earth Day achieved in 1970 helped lead to the creation of the United States Environmental Protection Agency (EPA), and the passage of the Clean Air, Clean Water and Endangered Species Acts (Earth Day Network 2018).

To continue with that momentum that started in 1970, a significant global moment came about with the creation of the United Nations Environment Programme (UNEP) in 1972 for the purpose of coherent implementation of sustainable development across the globe. Then, in 1974, UNEP started World Environment Day, a global platform to increase awareness.

It was not until 1987 that the defining of sustainability became well-documented. The United Nations' World Commission on Environment and Development (also referred to as the Brundtland Commission), stated that sustainability is economic development activity that "meets the needs of the present without compromising the ability of future generations to meet their own needs." The basic premise of sustainability is that Earth's resources cannot be used, depleted, or damaged indefinitely. Not only will these resources run out at some point, but their exploitation undermines the ability of life to persist and thrive (Portney 2015, 9). The Brundtland Commission described sustainability as having three equal parts, or three key elements starting with the letter "e," environment, economy, and equity. They were coined as the main pillars for creating a sustainable system (Portney 2015, 10).

The evolution and philosophy of sustainability has a unique history. The idea itself hasn't changed drastically, but the interpretation has changed often because of the many intellectuals, institutions, governments, and businesses having critiqued it over time. Every one of those interpretations of sustainability essentially agrees that humans and all living creatures and organisms on Earth will not survive if resources are not properly consumed and managed. Also, it is generally agreed that living in balance with nature needs to be taught and deeply nurtured. But how do we create peace, harmony, and balance with nature? Luckily, we finally have a significant amount of national and global movements, initiatives, policies, challenges, and business philosophies focusing on the sustainable development of Earth. Sustainability and green movements have become much more common over the last thirty years, and one of those movements was created in 1994 by John Elkington. Elkington coined the term and concept of "triple bottom line, or people, planet, profit" to create awareness and encourage organizations to ethically account for environmental impacts and social impacts while maximizing financial gains (Clark 2017). Other notable movements or developments accelerating sustainability for the planet since 1994 include the launch of the Global Reporting Initiative and standards in 1997, the United Nations Sustainable Development Goals in 2015, the Paris Climate Agreement in 2016, and COP26 United Nations Climate Change Conference in 2021. These are some of the major movements that have been created in response to climate change. What movement might happen next to help reduce waste and use it as a valuable resource to reduce human impact?

Step 1—Respect

Respect Earth and Understand Its Natural Systems

The Fast-Changing Landscape and Our Ways of Living

Before digging into what waste is, let's start with step 1, respect. This section takes an expedition into the history and evolution of how the United States, Michigan, and the world have arrived at their current lifestyles and why certain human forces have and are creating a fast-changing landscape.

All available resources come solely from Earth at this point, from no other place in the universe. We take what we need to survive, to thrive, and to live better lives. However, in the last 250-plus years, things have rapidly changed, starting with the Industrial Revolution (1750–1850), then moving into the Technological Revolution (1850–1950), and now the Digital Revolution (1950–current). Humans progressed and innovated from hand tools to machines, horse and wagons to fast cars and airplanes, typewriters to smart tablets, and from dreaming about walking on the moon to *literally* walking on the moon. Our nation and the world's technological advancements, brilliance, and ingenuity have accelerated the economic engine to extraordinary heights and unforeseen levels. Our world is now more interconnected and globalized than ever, and it looks like that momentum is going to continue. In Thomas L. Friedman's modern book, *Thank You for Being Late*, he talks about how the planet's three largest forces—Moore's law (technology), the market (globalization), and Mother Nature (climate change and biodiversity loss)—are accelerating simultaneously. Friedman's perspective is that we currently live in an age of many accelerations, and that we have the tools, resources, and opportunities to save the world or to destroy it quickly (Friedman 2016).

Not long ago, humans built things to survive and to meet basic needs. Some of these survival methods include shelters or dwellings, weapons and tools for killing and preparing animals to eat, wagons for hauling goods and people by horse, and wooden canoes to travel by water. Before the Industrial Age, human needs revolved more around survival, but they arguably lived tougher and shorter lives then. Ancient philosophy, from the likes of Aristotle, teaches that those basic human instincts and the purpose of life is to seek pleasure and to maximize happiness. Philosopher Jeremy Bentham preached the idea of utilitarianism, the goal of maximizing good for the greatest number of people. In the last few centuries, one could argue that those philosophical ideas and pursuits of happiness are being achieved. Thanks to humankind's pursuit of maximizing happiness and the good of people, amazing technological advancements have been generated. A person can now travel the planet faster and cheaper, indulge in more of its available resources, and start revolutionary businesses that provide new goods or services for humanity to enjoy. Simply put, because of these advancements, humans can consume more resources and live longer. Four key logistical hurdles that have been overcome through our major revolutions are:

1. The ability to better procure raw materials and resources *(resource extraction)*
2. Quick and efficient manufacturing *(mass production)*
3. Platforms to allow manufactured products and services to be available to more people at lower cost *(cost and convenience)*
4. The reduction in the time it takes a person to receive a good or service *(delivery and logistics)*

Overcoming these four logistical barriers has spurred growth in capitalism, business competition, and consumerism. It has given birth to our modern-day economic engine that provides seemingly unlimited resources that are available at the click of a button. The problem is, we don't live in a world of unlimited resources. In fact, we have a finite amount.

The Economy and Our Population

Going back to the founding principles of economics, we live in a world of scarcity, or at least we should. Scarcity or paucity, according to *Market Business News*, refers to limitation—limited supplies, components, raw materials, and goods—in an environment with unlimited human wants. It is the fundamental economic problem of having what appears to be limitless human wants in a world with limited resources. Determining how to make the best use of scarce resources is absolutely fundamental to economics. The factors of production are not limitless (i.e., there is scarcity). Therefore, we must make choices about how best to use them. This is where economics comes in. Factors of production are the building blocks or elements that we use to produce goods and services. Economists divide factors of production into land, capital, labor, and enterprise (entrepreneurship). In the world of economics, we must learn to live with one basic problem: the gap between scarcity—limited resources—and unlimited wants. To satisfy those wants, suppliers need to determine how to use those limited resources carefully (Market Business News 2018). A great question is whether our current human consumption of resources is in balance with what our Earth can provide; can the Earth regenerate its bioresources at a rate similar to our levels of consumption?

To offer a point of view on our current and forecasted consumption of resources, we can dig into gross domestic product (GDP), consumer spending, and population. Starting with GDP is generally the most common method to measure an economy's overall health and well-being. According to the *Merriam-Webster Dictionary's* financial definition, GDP is the broadest quantitative measure of a nation's total economic activity. More specifically, GDP represents the monetary value of all goods and services produced within a nation's geographic borders over a specified period of time (*Merriam-Webster* n.d.). To calculate GDP per capita, take total GDP and divide it by the current population. The basic formula equals the total consumption plus government expenditures, plus investment, plus exports, and minus imports. The World Bank national accounts historical data reports that in 1960, the GDP in the United States was $543.3 billion and grew to $19.4 trillion in 2017. The GDP has virtually grown year after year, except for three periods of major recession (1974–1975, 1982, and 2008–2009). Figure 1 shows the annual GDP percentage growth since 1995 and forecasted growth up to the year 2025. The forecast, by Deloitte Insights, predicts that the GDP will continue to increase by over 1 percent each year.

Figure 1—GDP Growth

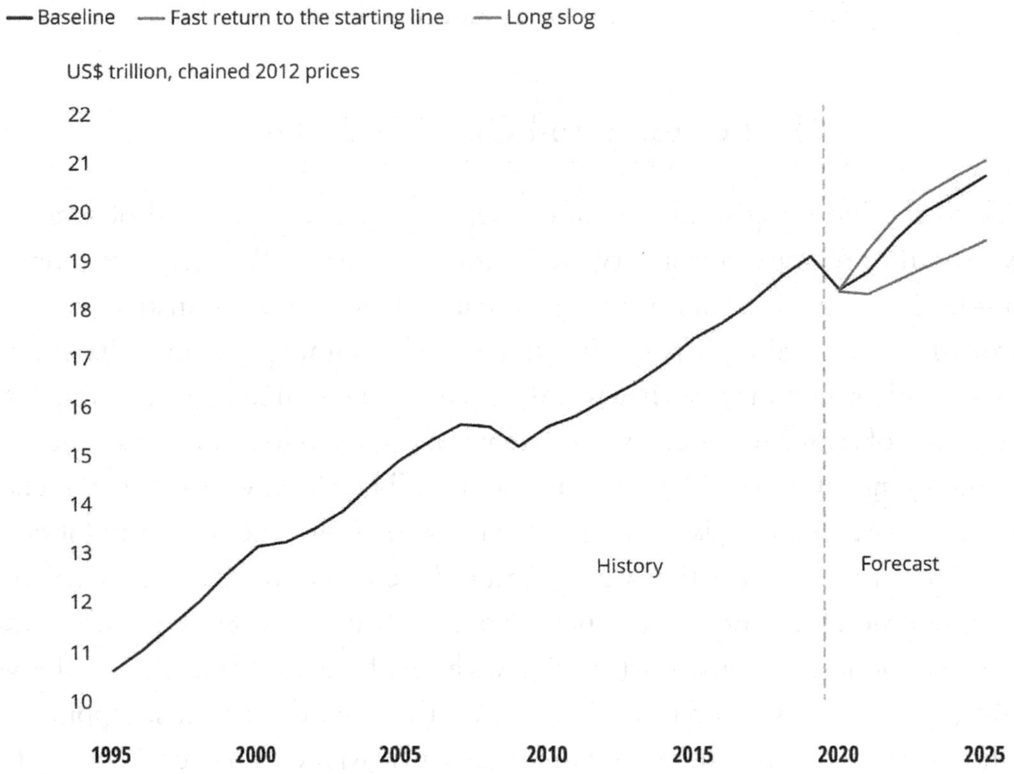

Source: Deloitte Insights, Deloitte Analysis.

Another indicator of our economic health is consumer spending. Consumer spending is the breakdown of the total volume of goods and services used or consumed per household. According to Investopedia, contemporary measures of consumer spending include all private purchases of durable goods (e.g., appliances), nondurables (e.g., food), and services (e.g., education and transportation) (Investopedia, n.d.). To put it simply, as consumer spending increases, more products and services are being purchased by people. Generally speaking, the more available funds people have, the more they might be willing to spend. Figure 2 shows the annual consumer spending percentage growth since 1995 and forecasted growth up to year 2025. The forecast, by Deloitte Insights, predicts that consumer spending will continue to increase a minimum of 1 percent each year through 2025.

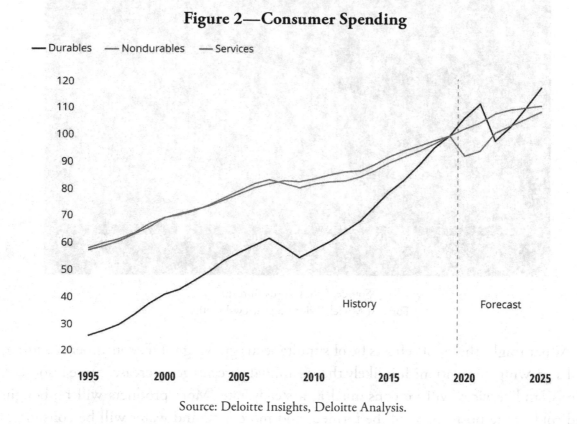

Figure 2—Consumer Spending

Source: Deloitte Insights, Deloitte Analysis.

The population of the United States has grown steadily since 1950, from 161 million to 328 million in 2018. According to the US Census Bureau (as of August 3, 2018), there is one birth every eight seconds, one death every twelve seconds, and one international migrant entry every twenty-nine seconds in America. That equals a net gain of one person every twelve seconds. However, globally, our population grows by roughly two people every second. Currently, our global population is almost 7.2 billion and is expected to grow to 9.6 billion by 2050 (United Nations Department of Economic and Social Affairs 2013). Many countries are becoming increasingly developed in terms of technology, resource extraction, and health care. And not only is the population growing, but people are living longer. With the population increasing both globally and on our own soil, the scarcity of resources will likely become a hot topic and, in many ways, already is, but I will delve into that further later. The US population is predicted to grow steadily and hit almost four hundred million by 2050. States like Arizona and Florida are among those forecasted to see larger growth. The population in Michigan has hovered around ten million since 2000 but is forecasted to steadily grow in the coming years. See Figure 3 below to visualize the trend of our population in the United States.

Figure 3—US Population

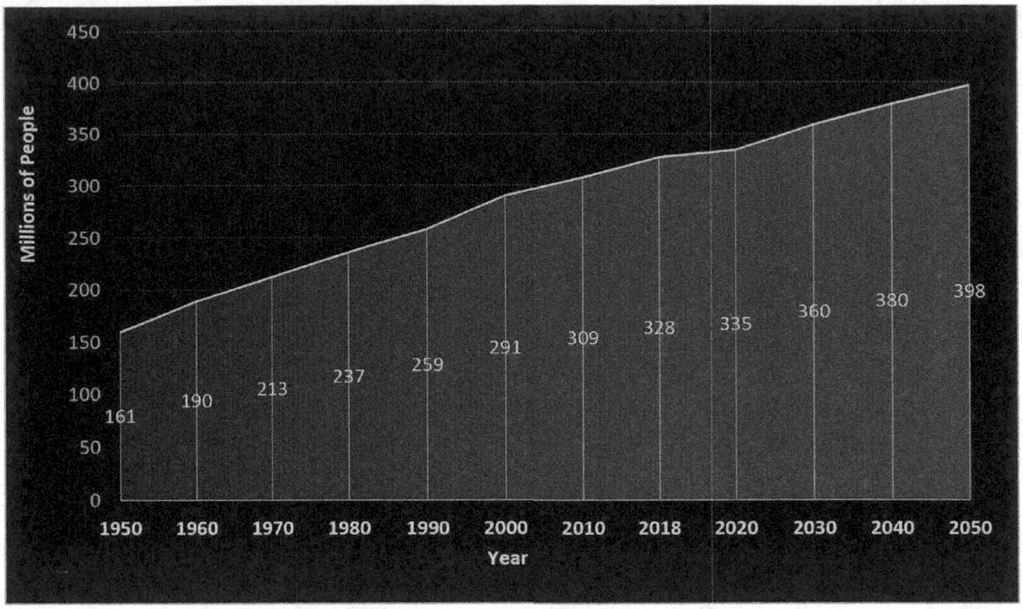

Source: U.S. Census Bureau.
Forecast Model: Tyler Kanczuzewski, 2018.

What might the ripple effects be of simultaneous growing GDP, consumer spending, and a growing population? It is likely that as annual percentages increase, more resources, goods, and services will be consumed at a steady rate. More products will be bought and sold, more businesses will be formed, and more food and water will be consumed. According to a US Small Business Administration 2018 report, there were 30.2 million small businesses, which employed almost 59 million people. The number of small businesses continues to grow year after year, and they employ 47.5 percent of the private workforce (US Small Business Administration 2018). It is important to note that these chosen indicators are not the only way to measure the health of and to forecast economic activity. However, if one would like to do research on the state of an economy and to grasp how things are trending, metrics like GDP, consumer spending, and population are essential. As these indicators grow, it becomes increasingly important for our population to visualize what the US biocapacity is, as well as our planet's overall biocapacity status. Let's take a dive into available biocapacity statistics, as well as our ecological footprints as a human race.

Ecological Footprint

There is a new calculation method to measure the consumption rate of resources by humans and animals, compared to how fast those resources can naturally regenerate; this method is called "ecological footprint." According to the Global Footprint Network, the ecological footprint is the only metric used to measure how much nature, or biodiversity, we have and how much nature we use (Global Footprint Network 2018).

> Defining ecological footprint:
>
> Ecological Footprint accounting measures the demand and supply of nature. On the demand side, the Ecological Footprint measures the ecological assets that a given population requires to produce the natural resources it consumes (including plant-based food and fiber products, livestock and fish products, timber and other forest products, space for urban infrastructure) and to absorb its waste, especially carbon emissions. The Ecological Footprint tracks the use of six categories of productive surface areas: cropland, grazing land, fishing grounds, built-up land, forest area, and carbon demand on land. On the supply side, a [city's, state's or nation's] biocapacity represents the productivity of its ecological assets (including cropland, grazing land, forest land, fishing grounds, and built-up land). These areas, especially if left unharvested, can also absorb much of the waste we generate, especially our carbon emissions.
>
> Both the Ecological Footprint and biocapacity are expressed in global hectares—globally comparable, standardized hectares with world average productivity. Each [city's, state's or nation's] Ecological Footprint can be compared to its biocapacity. If a population's Ecological Footprint exceeds the region's biocapacity, that region runs an ecological deficit. Its demand for the goods and services that its land and seas can provide—fruits and vegetables, meat, fish, wood, cotton for clothing, and carbon dioxide absorption—exceeds what the region's ecosystems can renew. A region in ecological deficit meets demand by importing, liquidating its own ecological assets (such as overfishing), and/or emitting carbon dioxide into the atmosphere. If a region's biocapacity exceeds its Ecological Footprint, it has an ecological reserve. Conceived in 1990 by Mathis Wackernagel and William Rees at the University of British Columbia, the Ecological Footprint launched the broader footprint movement, including the Carbon Footprint, and is now widely used by scientists, businesses, governments, individuals, and institutions working to monitor ecological resource use and advance sustainable development. (Global Footprint Network 2018)

The big question is about whether humanity is living in balance with the natural world. Are we respecting resources and the available biocapacities that Earth has provided for us and using them sustainably? The Global Footprint Network has analyzed the current world ecological footprint and has accumulated a large amount of data. The interpretation of this data shows the world operating at an ecological overshoot. According to the Global Footprint Network, since the 1970s, humanity has been in an ecological overshoot, with annual demand on resources exceeding what Earth can regenerate each year. Today, humanity consumes the equivalent to 1.7 Earths each year. This means it now takes Earth over one year and seven months to regenerate what is consumed in a year. The study shows we are using more ecological resources than nature can regenerate through activities like overfishing, overharvesting forests, and emitting more carbon dioxide into the atmosphere than our forests can sequester (Global Footprint Network 2018). See Figure 4 to visualize the significant change in footprint, especially starting after 1970.

Figure 4—World Ecological Footprint

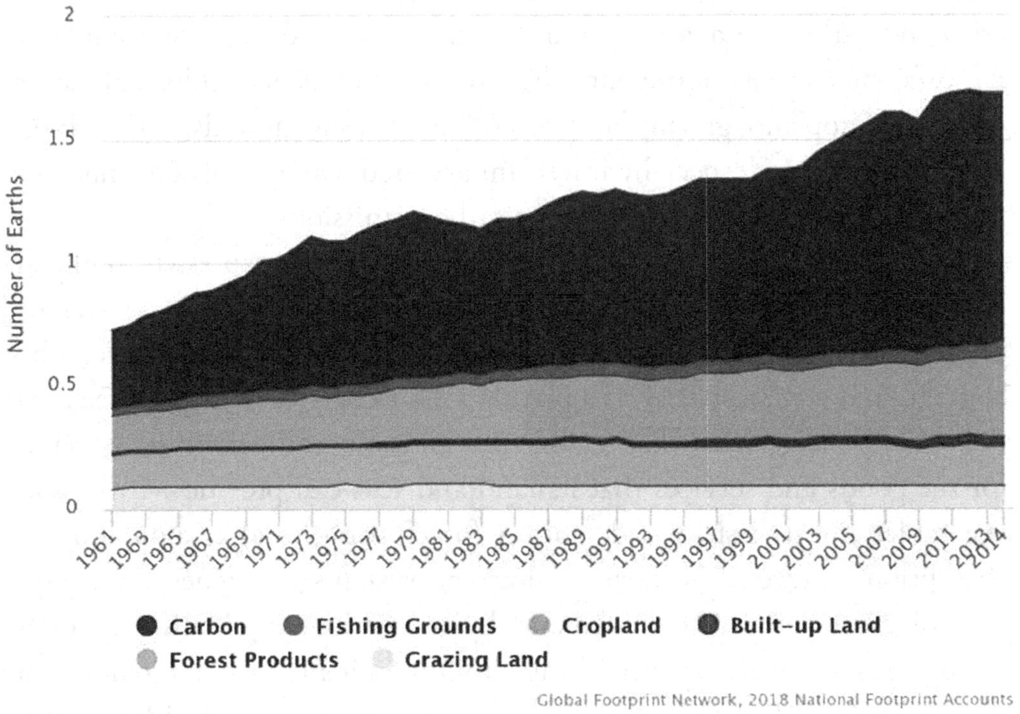

Global Footprint Network, 2018 National Footprint Accounts

Source: Global Footprint Network, National Footprint Accounts, 2018.

In the United States, the story is just as alarming. The old saying was that the United States has 5 percent of the world population but consumes 25 percent of the available resources. In reality, that saying is not far from the truth. Currently, the United States contains 4.4 percent of the global population and consumes 10–30 percent of the available global biocapacities. According to the Global Footprint Network, in 2014, the United States needed about five Earths to keep up with its human demand and was operating at 133 percent biocapacity deficit (see Figure 5). The ecological deficit poses a potential threat for the United States, and it will become increasingly more important for the country to understand how to minimize its deficit and strive to work more sustainably with nature.

Figure 5—US Ecological Footprint

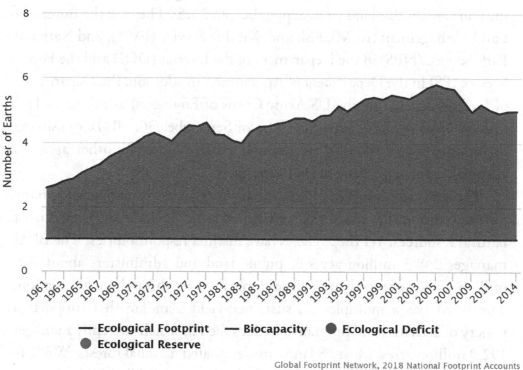

Source: Global Footprint Network, National Footprint Accounts, 2018.

The total land and water available in the United States, according to the US Department of Agriculture's Natural Resources Conservation Service (NRCS), is roughly 1.94 billion acres and includes the contiguous forty-eight states, Hawaii, Puerto Rico, and the US Virgin Islands (note that this number excludes Alaska). About 71 percent of the area is nonfederal rural land and nearly 1.4 billion acres. Nonfederal rural lands are predominantly forestland (413 million acres), rangeland (406 million acres), and cropland (363 million acres). See Figure 6, which shows a chart breakdown. Interestingly, if you include Alaska in that total land calculation for the United States, it adds 375 million acres, taking the total to almost 2.3 billion acres (Hull and Leask 2000, 1). However, Alaska is excluded in the NRCS reports because the federal government owns and manages 61 percent of the state. The NRCS provides updated information on the status, condition, and trends of land, soil, water, and related resources on the nation's nonfederal lands. Nonfederal lands also include privately owned lands, tribal and trust lands, and land controlled by state and local governments (US Department of Agriculture Natural Resources Conservation Service n.d.).

Federal land ownership, overview by the Congressional Research Service:

The federal government owns roughly 640 million acres, about 28 percent of the 2.27 billion acres of land in the United States (including Alaska). Four major federal land management agencies administer 610.1 million acres of this land (as of September 30, 2015). They are the Bureau of Land Management (BLM), Fish and Wildlife Service (FWS), and National Park Service (NPS) in the Department of the Interior (DOI) and the Forest Service (FS) in the Department of Agriculture. In addition, the Department of Defense (excluding the US Army Corps of Engineers) administers 11.4 million acres in the United States (as of September 30, 2014), consisting of military bases, training ranges, and more. Numerous other agencies administer the remaining federal acreage.

The lands administered by the four major agencies are managed for many purposes, primarily related to preservation, recreation, and development of natural resources. Yet the agencies have distinct responsibilities. The BLM manages 248.3 million acres of public land and administers about 700 million acres of federal subsurface mineral estate throughout the nation. The BLM has a multiple-use, sustained-yield mandate that supports a variety of activities and programs, as does the FS, which currently manages 192.9 million acres. Most FS lands are designated national forests. Wildfire protection is increasingly important for both agencies. The FWS manages 89.1 million acres of the US total, primarily to conserve and protect animals and plants. The National Wildlife Refuge System includes wildlife refuges, waterfowl production areas, and wildlife coordination units. In 2015, the NPS managed 79.8 million acres in 408 diverse units to conserve lands and resources and make them available for public use. Activities that harvest or remove resources from NPS lands generally are prohibited.

The amount and percentage of federally owned land in each state varies widely, ranging from 0.3 percent of land (in Connecticut and Iowa) to 79.6 percent of land (in Nevada). However, federal land ownership generally is concentrated in the West. Specifically, 61.3 percent of Alaska is federally owned, as is 46.4 percent of the 11 coterminous western states. By contrast, the federal government owns 4.2 percent of lands in the other states. This western concentration has contributed to a higher degree of controversy over federal land ownership and use in that part of the country.

Throughout America's history, federal land laws have reflected two visions: keeping some lands in federal ownership while disposing of others. From the earliest days, there has been conflict between these two visions.

During the 19ᵗʰ century, many laws encouraged settlement of the West through federal land disposal. Mostly in the 20ᵗʰ century, emphasis shifted to retention of federal lands. Congress has provided varying land acquisition and disposal authorities to the agencies, ranging from restricted (NPS) to broad (BLM). As a result of acquisitions and disposals, from 1990 to 2015, total federal land ownership by the five agencies declined by 25.4 million acres (3.9 percent), from 646.9 million acres to 621.5 million acres. Much of the decline is attributable to BLM land disposals in Alaska and to reductions in DOD land. By contrast, land ownership by the NPS, FWS, and FS increased over the twenty-five-year period. Further, although 15 states had decreases of federal land during this period, the other states had varying increases.

Numerous issues affecting federal land management are before Congress. These issues include the extent of federal ownership and whether to decrease, maintain, or increase the amount of federal holdings; the condition of currently owned federal infrastructure and lands and the priority of their maintenance versus new acquisitions; and the optimal balance between land use and protection, and whether federal lands should be managed primarily to benefit the nation as a whole or to benefit the localities and states. Another issue is border control on federal lands along the southwestern border, which presents challenges due to the length of the border, remoteness and topography of the lands, and differences in missions of managing agencies." (Congressional Research Service 2017, 2)

Just as resources are scarce, so too is land. The United States governs roughly 2.3 billion of the 36.7 billion acres, or 6.3 percent, of the total land on Earth. The total land surface area of Earth is about 57,308,738 square miles, of which about 33 percent is desert and about 24 percent is mountainous. Two examples of uninhabitable land are the continent of Antarctica and Death Valley, California. Geographic regions like mountains, deserts, and swamplands are basically uninhabitable. Subtracting this uninhabitable 57 percent (32,665,981 mi²) from the total land area leaves 24,642,757 square miles or 15.77 billion acres of habitable land (Pianka n.d.). The world population currently exceeds 7.5 billion, so if people were spread evenly across the globe to inhabit the usable land, each person could occupy almost half an acre. The rest of planet is made up of water. The oceans make up 97 percent of the total water and 70 percent of the planet's surface area (National Ocean Service n.d.). The United States is made up of mostly habitable land, so of the estimated 15.77 billion acres of habitable land on Earth, the United States comprises almost 10 percent, which is interesting, as the United States makes up only 4.4 percent of the current global population as of this writing.

Land will become increasingly valuable and scarce as the population continues to rise. In the coming years, governments, organizations, businesses, and citizens of Earth will likely be challenged more than ever to come up with strategies and solutions to protect Earth and its valuable resources. World government organizations like the United Nations have already created awareness movements in the last few decades to fight climate threats such as global warming and pollution. Movements can be successful if proper standards are established and goals and metrics are put in place to measure progress. The United States has a great opportunity to step up as a steward and protect nature. The federal government demonstrates forms of leadership when it comes to protecting federal land, as previously referenced by the Congressional Research Service. The federal government and other organizations can be constructive watchdogs and protectors of land, especially if standards and codes of ethics are put in place and followed routinely. For example, National Park Service–protected land is prohibited from being touched and can be recreationally used only to certain extents. If similar protection methodologies could be adopted to protect the rest of US lands, would resources and biocapacities be better managed to meet the needs of future generations?

Figure 6—US Natural Resources Inventory Report, 2012 (Excludes Alaska)

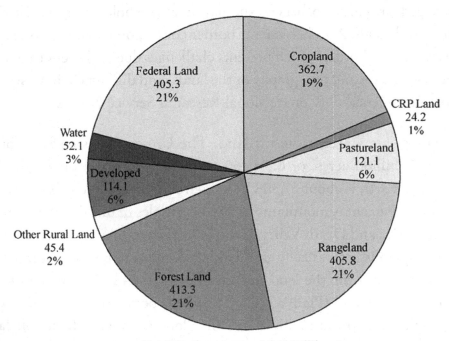

Total Surface Area = 1,944 Million Acres
(Cropland includes cultivated and noncultivated cropland)

Source: US Department of Agriculture Natural Resources Conservation
Service, 2012 National Resources Inventory Report.

Human Development Index

The United States consumes much of the world's resources, as data shows, but why? Logic and basic reasoning would point to metrics such as population, GDP, consumer spending, various economic indicators, income per capita, education, and health care. A way to measure those particular variables is through human development, or the human development index (HDI).

According to the United Nations Development Programme (UNDP):

> Human development is about acquiring more capabilities and enjoying more opportunities to use those capabilities. With more capabilities and opportunities, people have more choices, and expanding choices is at the core of the human development approach. But human development is also a process. Anchored in human rights, it is linked to human security. Its ultimate objective is to enlarge human freedoms. Human development is development of the people through the building of human resources, for the people through the translation of development benefits in their lives and by the people through active participation in the processes that influence and shape their lives. Income is a means to human development but not an end in itself. (United Nations Development Programme n.d.)

Figure 7 shows a circular flow of capabilities and opportunities. The significance of the figure is to illustrate how it takes a compilation of key variables to create an environment that is enticing for humans to live, thrive, and develop in a particular geographic area.

Figure 7—Human Development for Everyone

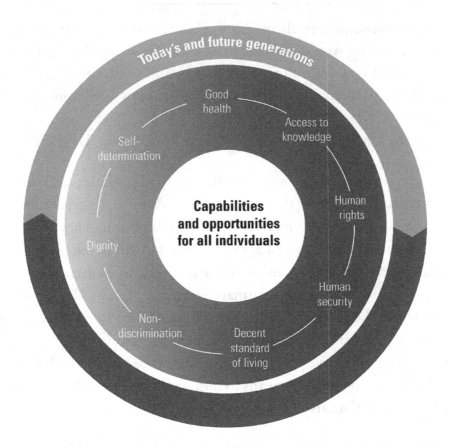

Source: UNDP. 2016. Human Development Report 2016: Human Development for Everyone. New York, http://hdr.undp.org/en/content/human-development-report-2016.

The Human Development Index was created in 1990 for the purpose of assessing achievements in the basic dimensions of human development, as well to compare and rank countries by their ability to sustain and grow human development measures. According to the UNDP, those dimensions of human development are: to lead a long and healthy life, measured by life expectancy at birth; to acquire knowledge, measured by mean and expected years of schooling; and to achieve a decent standard of living, measured by gross national income per capita (United Nations Development Programme n.d.). According to the 2016 index, the United States was ranked tenth in the world. The US population is much larger than all other countries that comprise the top ten. See Figure 8 to visualize how the United States compares to the other countries in the top ten. These countries, including the United States, are listed as very high in human development.

Figure 8—Nations with Highest HDI Index

HDI rank		Human Development Index (HDI)	Life expectancy at birth	Expected years of schooling	Mean years of schooling	Gross national income (GNI) per capita	GNI per capita rank minus HDI rank	HDI rank
		Value	(years)	(years)	(years)	(2011 PPP $)		
		2015	2015	2015[a]	2015[a]	2015	2015	2014
VERY HIGH HUMAN DEVELOPMENT								
1	Norway	0.949	81.7	17.7	12.7	67,614	5	1
2	Australia	0.939	82.5	20.4[b]	13.2	42,822	19	3
2	Switzerland	0.939	83.1	16.0	13.4	56,364	7	2
4	Germany	0.926	81.1	17.1	13.2[c]	45,000	13	4
5	Denmark	0.925	80.4	19.2[b]	12.7	44,519	13	6
5	Singapore	0.925	83.2	15.4[d]	11.6	78,162[e]	−3	4
7	Netherlands	0.924	81.7	18.1[b]	11.9	46,326	8	6
8	Ireland	0.923	81.1	18.6[b]	12.3	43,798	11	8
9	Iceland	0.921	82.7	19.0[b]	12.2[c]	37,065	20	9
10	Canada	0.920	82.2	16.3	13.1[f]	42,582	12	9
10	United States	0.920	79.2	16.5	13.2	53,245	1	11

Source: UNDP. 2016. Human Development Report 2016: Human Development for Everyone. New York, http://hdr.undp.org/en/content/human-development-report-2016.

Figure 9 shows the same countries and their ranks, and rates them based on environmental sustainability, economic sustainability, and social responsibility.

Figure 9—Sustainability Ratings

Country groupings (terciles)

Top third	Middle third	Bottom third

Three-colour coding is used to visualize partial grouping of countries by indicator. For each indicator countries are divided into three groups of approximately equal size (terciles): the top third, the middle third and the bottom third. See *Notes* after the table.

HDI rank		Environmental sustainability							Economic sustainability					Social sustainability			
		Renewable energy consumption	Carbon dioxide emissions		Forest area		Fresh water withdrawals	Natural resource depletion	Adjusted net savings	External debt stock	Research and development expenditure	Concentration index (exports)	Income quintile ratio	Gender Inequality Index	Population in multi-dimensional poverty	Old-age (ages 65 and older) dependency ratio
		(% of total final energy consumption)	Per capita (tonnes)	Average annual change (%)	(% of total land area[a])	Change (%)	(% of total renewable water resources)	(% of GNI)	(% of GNI)	(% of GNI)	(% of GDP)	(value)	Average annual change (%)	Average annual change (%)	Average annual change (%)	(per 100 people ages 15–64)
		2012[b]	2013	1990/ 2013	2015	1990– 2015	2005–2014[c]	2010– 2014[c]	2005– 2014[c]	2005– 2014[c]	2005–2014[c]	2014	2000/2014	2005/2015	2005/2014	2030[d]
VERY HIGH HUMAN DEVELOPMENT																
1	Norway	58.0	11.7	2.0	33.2	−0.2	0.8	7.1	21.0	..	1.7	0.372	..	−3.8	..	32.2
2	Australia	8.4	16.3	0.2	16.2	−2.9	3.9	3.3	8.5	..	2.2	0.266	0.0	−1.4	..	31.3
2	Switzerland	22.7	5.0	−1.0	31.7	9.0	3.8	0.0	15.0	..	3.0	0.256	..	−5.1	..	38.3
4	Germany	12.4	9.2	..	32.8	1.2	21.4	0.0	13.3	..	2.9	0.097	..	−4.3	..	47.7
5	Denmark	27.6	6.8	−1.8	14.4	12.6	10.6	1.0	14.5	..	3.1	0.086	..	−3.7	..	37.1
5	Singapore	0.5	9.4	−2.1	23.1	−5.2	..	0.0	36.9	..	2.2	0.250	..	−5.3	..	36.5
7	Netherlands	4.7	10.1	−0.2	11.2	9.3	11.8	0.4	16.9	..	2.0	0.097	..	−4.3	..	41.9
8	Ireland	7.0	7.6	−0.7	10.9	67.2	1.5	0.1	16.1	..	1.5	0.241	..	−3.5	..	29.2
9	Iceland	78.1	6.1	−1.1	0.5	205.6	1.8	0.0	11.2	..	1.9	0.445	..	−5.9	..	32.5
10	Canada	20.6	13.5	−0.6	38.2	−0.3	1.3	2.1	7.0	..	1.6	0.179	−0.2	−3.1	..	38.5
10	United States	7.9	16.4	−0.7	33.9	2.7	13.6	0.7	6.4	..	2.7	0.095	0.4	−2.8	..	33.8

Source: UNDP. 2016. Human Development Report 2016: Human Development for Everyone. New York, http://hdr.undp.org/en/content/human-development-report-2016.

Pictured Rocks National Lakeshore, Michigan.

Ecological Assessment of Michigan

The geography of the United States is vast and diverse, ranging from habitable to uninhabitable, desert to swampland, mountains to plains, farmland to developed urban environments, and dense forests to rangelands. Each state and geographic region is unique. Some states have more lakes, rivers, or ocean shorelines. More water should equate to having more fish to eat, or, potentially, more water for drinking. Some states have land that is better suitable for agriculture or environments that can grow certain fruits and vegetables compared to others that cannot. Some regions have dense forests, which can be logged and turned into wood for various uses. Certain regions contain more wildlife because of the natural terrain and abundance of food and water. Various regions even have certain minerals or nonrenewable fossil-fuel resources like coal, oil or natural gas reserved deep in the ground.

Each state or geographic region has different ecological reserves and resources available to its inhabitants. Local and state governments, the federal government, associations, businesses, and constituents typically have the right to manage and use these resources to their benefit. The benefits may range from economic prosperity and opportunities to the health and well-being of people, to the protection and security of nature and its wildlife, or for various recreational uses. The questions become, what amounts of ecological reserves can be consumed, what should be protected, and what ethics and standards ought to be followed? All are imperative questions and ongoing dilemmas regarding ecological reserves and resources across the country.

Evidence shows that some states and regions are maintaining immensely different ecological reserves and deficits. Some resources are thought to be more valuable than others, and therefore are consumed or extracted in higher volumes. The root cause of why resources are being consumed at extreme rates is likely the rapid advancement of technology throughout the recent industrial, technological, and digital revolutions—for example, higher-tech machinery and developing innovative computer technology. The growing economic engine, population, human development, and increase in goods and services traded globally also play a significant role in the growth of resource consumption. In general, the more goods that are consumed by people, the more biocapacity is needed.

When it comes to an abundance of nature, the state of Michigan sits in what some may call a geographic honey hole. Nature can provide resources that are either renewable or nonrenewable. Some examples of renewable resources are trees, plants, and animals (flora and fauna). There are also renewable energy resources, or forces of nature, that occur naturally on Earth every day (e.g., solar, wind, hydro, tidal, and geothermal). As far as humans know, those forces of nature will occur abundantly as long as Earth exists. Unfortunately for Earth, the sun is predicted to turn into a red giant and swallow up the Earth in billions of years, leading to the end of Earth's existence (BBC Science Focus 2022). Nonrenewable resources range from minerals, fossil fuels, soil, and more.

Michigan, in general, has a vast amount of both renewable and nonrenewable valuable resources, but some of those resources have already been entirely mined, cut, pumped, or extracted in some shape or form. This honey hole that Michigan sits in may have happened by chance, by natural forces, or by major events. No one knows for certain. Historical records and new research shines light on what has happened with some of those valuable resources throughout time and how they have evolved. Has the state of Michigan managed its resources and biocapacities properly to sustain life and opportunity for its current population as well as future generations?

Pictured Rocks National Lakeshore, Michigan.

Eastern White Pine Tree.

Timber talk: In the 1800s, the Michigan logging and deforestation industry was one of the main economic drivers for the state. Michigan was saturated with eastern white pines, which is the state tree. Those white pines, along with other hardwood trees, were almost completely logged and sent to mills and waterways to be used across the Midwest, the United States, and the world. The abundance of waterways allowed for logging companies to easily transport the highly valued lumber. The city of Chicago had a massive fire in 1871, and it is recorded that Michigan lumber helped rebuild the city (Husain 2021). There are still some white pines across the state in remote areas, such as Hartwick Pines State Park and other state or nationally protected forest regions. But if it wasn't for protection, would there be any white pines left? White pines are some of the bigger conifer trees in the country, inhabiting northern regions of the Midwest and Northeast. They provide protection and habitats for many birds, insects, and various animals. They are also known to be highly efficient in sequestrating large amounts of greenhouse gases (GHGs), particularly carbon dioxide (Seymour 2016).

Michigan's natural landscape and ecosystem have changed drastically because of logging. If you ask the typical economist about the Michigan logging industry, he or she would likely tell you it helped Michigan to become a very prosperous and populated state. Ask a conservationist, and he or she might express concern about how the flora and fauna have been negatively impacted. During 2017, approximately 60,012 acres of state forestland were prepared for timber harvests that yielded more than 1.1 million cords of wood, and the state received fifty million dollars in timber revenue. The forest products industry contributes $21.2 billion per year to Michigan's economy, supports ninety-nine thousand jobs, and creates one-third of the manufacturing jobs in the Upper Peninsula (Michigan Department of Natural Resources 2018).

The outlook for Michigan forestry shows some promise. According to the Michigan Department of Natural Resources (DNR), forest acreage has remained relatively stable since the 1950s, at approximately 19.3 million acres. Losses or conversions of forestland have tended to be compensated for by other lands being converted into forestland. The predominant land type converting into forestland has been agricultural land. In contrast to the stable forest acreage, total standing timber volumes have tripled since the middle of the last century, reflecting a maturing forest. The expanding volume also indicates that more growth has been continuously added to the forest than what has been removed or has died through natural causes, as evidenced by annual growth that has increased over the past fifty years. Michigan's surplus growing stock (annual net growth less harvests) is among the largest in the nation, with forests currently growing more than twice as much wood than is being harvested each year. This trend is expected to continue (Northeast-Midwest State Foresters Alliance n.d.). According to the Michigan DNR, the state's forest management practices meet the highest standards for environmental and social benefits. It was among sixteen organizations and individuals in 2017 to earn Leadership Awards from the Forest Stewardship Council (FSC), the world's leading forest certification system. The DNR is one of the longest-standing FSC-certified forest managers in the lake states region. FSC certification involves a rigorous, independent review of forest management. Michigan is certified by both the FSC and the Sustainable Forestry Initiative (Michigan Department of Natural Resources 2018).

The Great Lakes Basin.

Water talk: Michigan is surrounded by three of the five Great Lakes, and together, all five of the lakes and their watershed systems (Great Lakes Basin) make up roughly 20 percent of the world's usable fresh water. Michigan can give thanks to the Canadian and Arctic glaciers that stopped carving land and slowly melted in the Midwest tens of thousands of years ago (Egan 2017, 36). Michigan has a myriad of ponds, marshes, lakes, streams, rivers, and aquifers. After Alaska, Michigan has the most miles of shoreline in the United States Waterways make for a convenient method to ship goods and transport people, and Michigan's industries related to water have been another huge economic driver. The Great Lakes provide a booming economic engine. Some of those industries that benefit from the abundant water source are agriculture, fishing, shipping, transportation, recreation, and tourism, electric power plants, manufacturers, and various technology businesses.

It is estimated that the Great Lakes provide trillions of dollars in revenue to the economic system (Great Lakes Commission n.d). However, there are always pros and cons with a thriving source of economic revenue. Studies have shown that some of these industries have and currently pollute the water or have brought invasive species to the Great Lakes' aquatic ecosystem. European zebra mussels are a good example of how invasion can happen quickly. Zebra mussels attach themselves to the bottom of industrial ocean freighters in European seas to eat certain algae for nourishment. Those ships travel into the Great Lakes through the manmade water locks of the St. Lawrence Seaway in order to transport or pick up goods for trade. The zebra mussels detach themselves from the freighters upon arriving at the Great Lakes. Why? The Great Lakes provide new, unique, and fresh algae for zebra mussels to consume. Thus, zebra mussels have helped kill off various natural aquatic inhabitants throughout the Great Lakes and have shifted things out of balance (Egan 2017, 36). Scientist are studying the potential ramifications of invasive species like the zebra mussels.

How far out of balance things have become, or potentially will become, is up for debate. The condition of drinking water in Michigan is another potential issue. Misfortunes like the Flint water crisis can paint a picture of how water infrastructures can collapse and become highly contaminated. Power plants, such as nuclear, and industrial plants, such as iron and steel, use water from the Great Lakes every day. Most often, these plants use water for cooling methods and recycle the water back into the Great Lakes, supposedly at a safe and sanitary level. However, it is likely that the temperature of the recycled water is hotter than average and potentially affecting the lakes' natural temperatures.

Some industries have been given approval by federal and state governments to do controlled dumping of various types of waste directly into waterways. However, controlled dumping can become dangerous and controversial if systems are not well-managed (Pete 2017). The Great Lakes Basin provides usable water to a large portion of Midwest inhabitants and wildlife, including aquatic animals and insects. Earth's creatures can't survive without clean water. The world is getting warmer, and water is evaporating exponentially; simultaneously, the population is growing and people are living longer. Thus, it is highly likely there will be a significant increase in demand for clean and fresh water.

In the United States, there is already conversation about piping water from the Great Lakes to the Southwest, which is the driest part of the country and is already dealing with water shortages (Huttner 2015). Local and state governments, such as the city of Chicago or city of Grand Rapids, and various businesses like Nestlé, get approval to extract water from the Great Lakes for drinking water or for business use. Water extraction is delicate, and the proper infrastructure is required to safely and properly manage it. In some situations, water extraction approval is obtained via the International Joint Commission (IJC), an international organization created and signed by Canada and the United States. The commission typically manages and monitors how much water may be pumped daily from the Great Lakes water system. The IJC is a critical agency because both Canada and United States surround the Great Lakes, and both countries are also responsible for protecting them. Extracting and cycling water back into the Great Lakes Basin is a delicate process and must be constantly monitored to ensure water levels are maintained and that water quality meets agreed-upon standards. Infrastructure such as wastewater treatment and recovery plants, as well as water transportation pipes, are examples of systems that require intensive daily monitoring and maintenance to meet quality and sanitary standards. Most of the water quality standards were created and are now monitored by the EPA.

Minerals and fossil fuel talk: Michigan's copper, iron ore, natural gas and oil stocks were once prevalent. In the 1800s, along with the logging industry, Michigan was a prime destination for companies to build operations to exploit mineral resources such as iron ore or copper. Michigan was one of the largest suppliers of copper and iron ore in the Midwest and in the United States during the Industrial Revolution. These industries have slowed down immensely, and most of the copper and iron ore have been mined completely in Michigan. The impacts from mining these particular resources are waste and pollution, land degradation, deforestation, water pollution, and acid mine drainage (GreenSpec, n.d.). Natural gas and oil production, or fossil fuel extraction, have been ongoing in Michigan for over a century. The most notable shale is the Antrim in northern Michigan. Michigan was the sixteenth-largest oil producer and thirteenth-largest natural gas producer in the United States in 1994 and was top among Midwestern states (MSU Department of Geography and College of of Social Science n.d.). Because of technology, industries have been able to more efficiently extract natural gas in the last twenty years or so. The problem with these new technologies is that they use methods that are potentially harmful to Earth. Some of these methods, like fracking, use various forms of harmful chemicals that are dispersed deep into the ground, where they potentially seep into vital water aqueducts used by society. Other methods are ground removal, or explosion, to access portions of the shale that are richer with gas. The ramifications through these methods are being closely studied.

Agriculture talk: In more recent years, Michigan's agriculture development has become a key source of revenue generation. In fact, Michigan's food and agriculture sector generated more than a hundred billion dollars of total economic activity in 2017, and 923,000 people were employed by the sector, which accounts for about 22 percent of the state's employment (Michigan Department of Agriculture and Rural Development 2017). According to the Michigan Department of Agriculture, Michigan is one of the top states in the United States when it comes to the diversity of agricultural products grown with significant value-add opportunities in agriculture, food, and forest products. Also, Michigan's central location is five hundred miles of almost 50 percent of the North American population and has a network in place to meet the needs of those populations (Michigan Department of Agriculture and Rural Development 2017). According to the 2012 USDA Census of Agriculture report, Michigan had 52,194 farms, equaling 9.95 million acres of total land (see Figure 10). Michigan's total surface area is thirty-seven million acres, so farmland equals almost one-third of available land surface. How does large-scale farming affect Michigan? Take, for example, water. Billions of gallons are consumed annually for watering crops and providing drinking water for livestock. Is the water being depleted or being contaminated before it's recycled back into the ecosystem? Those questions are constantly being analyzed and measured by the environmental stewardship division of the Michigan Department of Agriculture and Rural Development (Michigan Department of Agriculture and Rural Development 2017).

Figure 10—Census of Agriculture, 2012, Michigan

All farms		2012	2007	2002	1997
Farms ....number		52,194	56,014	53,315	53,519
Land in farms ....acres		9,948,564	10,031,807	10,142,958	10,443,935
Average size of farm ....acres		191	179	190	195
Estimated market value of land and buildings [1]:					
Average per farm ....dollars		766,148	610,556	509,299	335,580
Average per acre ....dollars		4,020	3,409	2,667	1,704

Source: USDA, National Agricultural Statistics Service.

In Michigan, and across the United States, people tend to cluster together to form urban centers, which typically create developed areas. People typically cluster to these areas because of economic opportunity, convenience, and for the comfort and security a developed area provides. Some examples of comfort and security are: easy access to biocapacity and goods and services; being a central or logistical location; and an abundance of food, water, neighborhoods, and recreational opportunities. It is interesting to understand how a state becomes developed (see, for example, Figure 11, a map of Michigan landcover classification for 2000). The southern section of the state happens to sustain the highest concentration of agriculture and urban development. How sustainable is Michigan in terms of managing its landcover and its valuable resources? It's an important question, and one that needs ongoing monitoring because things are always evolving. Business goes on, and people make decisions every day that can affect nature positively or negatively.

Figure 11—Michigan Landcover Classification, 2000

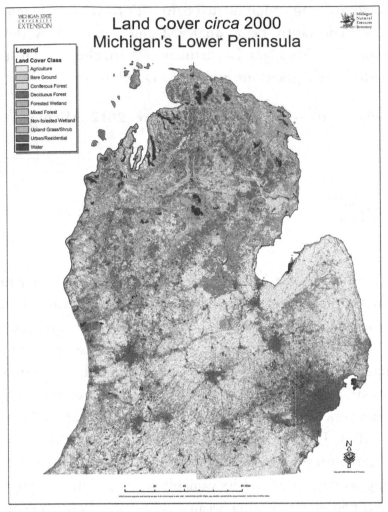

Source: Michigan DNR 2001, Michigan Forest Resource Assessment & Strategy
Report: http://169.62.82.226/documents/dnr/Strategic_457570_7.pdf.

Respectful Awareness of the Natural World

The goal of step 1, respect, is to raise awareness for the natural world in which we live. We live with scarcity, where the human race has boundless wants in a world of limited resources and supplies. Our pursuit of happiness and maximizing opportunity is not necessarily a bad thing, but we should understand that every decision we make has consequences, both positive and negative. How do we strive to live in balance with the natural world?

Do you ever think the world is accelerating fast or that we are moving into uncharted waters, figuratively speaking? Our major revolutions of change in the United States (industrial, technological, and digital) have certainly accelerated human development. Advancement in technology through our revolutions has created a modern world of growing brilliance, new inventions, and new businesses. Our trek into the future to develop and improve has allowed us to overcome many logistical barriers that, in many ways, are making life easier. As mentioned earlier, four key logistical hurdles that have been overcome through our major revolutions are resource extraction, mass production, cost and convenience, and delivery. This movement of overcoming logistical barriers has spurred growth in capitalism, business competition, and consumerism. This movement has also given birth to our modern-day globalized economic engine, an engine that provides seemingly unlimited resources that are available at the click of a button. But as we now know, we do not live in a world of unlimited resources!

Is Thomas L. Friedman correct in that technology, globalization, climate change, and biodiversity loss are all accelerating simultaneously? The data and research in this report shows that he is accurate. Contributing factors (e.g., GDP, consumer spending, and population), are persistently growing. Our human development in the United States and across the globe is growing, as we found in the Human Development Index. As research shows, the United States and the rest of the world are already running ecological deficits. Our biocapacities do not have enough time to regenerate at the rate we are consuming. Is our US and global population growing too quickly and going to deplete food and water availability in the near future? Is our potential worst enemy the fact that business is booming fast, and the economy is thriving? Are we too afraid of the consequences of slowing down or stabilizing the economy? Or do we continue at the current rate, but learn how to wisely consume and squeeze as much as we can from the resources?

Our advancement in technology and science is amazing, no doubt. Human ingenuity has given us the capability to innovatively blend and build things not natural to Earth (e.g., industrial plastics and rubbers, cardboard, and various chemicals). Those are just a few examples, and all were created to serve a purpose of convenience. We now make amazing products like vehicles, buildings, roads, and plastics. We did not just pull these things from the ground; instead, it has taken advanced science and human ingenuity to manufacture them. Products can make life better, but we must nurture the natural world and build things with nature in mind.

Products are being manufactured and consumed exponentially and are expected to continue as the world continues to develop and become more populated. There is a serious problem, and one that step 2, recover, addresses. Consuming more biocapacities and Earth's resources has given birth to this byproduct we call waste, and there is a lot of it on the planet right now. There are many types of waste, and this study will dig into that topic and focus more on one specific type of waste—product and packaging (material) waste. For as long as humans inhabit Earth, our activities will create certain types of waste. Fortunately, we have the choice to be stewards and do something about it. If we act as courageous stewards, we can make the best out of a potentially dangerous path.

Tips for Stewards: The Big Six for Respecting Nature

1. Respect the Natural World and Educate Yourself

    ☐ This can actually be fun. Let your curiosity about the world roam free. Go out and adventure; seek the world's beauty. While you're doing that, learn about the health of the flora and fauna, as well as land, air, and water health. See how humankind is impacting the natural world and its ancient natural systems. Most importantly, respect the natural world and try to live in balance with it.

    ☐ Research and learn about our ecology, and don't stop. Nature is always changing, and the world is constantly evolving, so we must continue learning about nature and how humans continually impact it in terms of biocapacity and resource consumption.

2. Protect Nature

    ☐ Stand up for what you believe is right. Be cautious of every decision you make and how it affects nature and the world around you, because we only have one planet. Treat the planet how you would want to be treated.

    ☐ Inform others, raise awareness in your local community, or create movements to protect the natural world. Start small but think big!

    ☐ Help governments or organizations enforce policies that will restrict or set limits on extraction of natural resources and biocapacities.

    ☐ Start a company, organization, or movement that protects nature.

    ☐ Challenge yourself and others to be ambassadors for positive change.

3. Adopt a Sustainable Mindset: "Nature First"

    ☐ Think sustainably every day, with every decision you make. Basically, in whatever you do or plan to do, think about how that decision will affect future generations' abilities to meet their needs. Only take or make what you need to survive and thrive. Most Native American tribes think seven generations ahead.

    ☐ Practice the "Leave No Trace Seven Principles" for outdoor ethics, and for practicing very low-impact activities on nature and in life (Marion 2014).

    ☐ Think before you have a child, build a building, make a product, or consume a good or service. How will that decision affect available biocapacities and resources, the environment, and future generations?

    ☐ Take advantage of and consume more renewable resources (e.g., solar, wind, geothermal, tidal).

☐ Use sustainably sourced or practiced, locally made or grown, organic, and more abundant materials (e.g., bamboo and hemp).

☐ Create a pro-versus-cons breakdown list for any key life decision. Make sure the pros always outweigh the cons and that pros put nature in mind first.

4. Reduce Consumption and Think Conservatively

☐ Consume only what you need to survive and thrive; in other words, reduce your consumption of products, goods, services, biocapacities, and Earth's valuable resources.

☐ Be efficient with your use of resources and maximize their values. Reuse or repurpose things when possible.

☐ By reducing consumption, we lower our own personal ecological footprints on Earth. We also pressure businesses and organizations to innovate and create new platforms that are more sustainable and more focused on nature.

5. Create Cohesiveness between Science and Technology and the Artificial World versus the Natural World

☐ Use science and technology to learn about nature and to measure and monitor key data. Use that data to make critical decisions.

☐ Use science and technology to support and help nature thrive.

☐ Create businesses and technology that mimic or respect the natural world.

☐ If you're operating a business or are a manufacturer, strive to build products that are eco-friendly or that use fewer resources or materials. More will be covered on this topic in steps 2 and 3.

6. Accept Consequences, Take Responsibility and Act; Be a Champion for Change

☐ Understand that every decision has a consequence potentially positive or negative. If the outcome to your activity is negative, take responsibility to correct it and act quickly to reverse or reduce the negative results.

☐ Be a champion for change in your local community and the world. Do not give up; continue fighting to save your family, your business or organization, your policies you helped enforce, your local community, and your planet. After all, we have only one planet; always remember that!

Step 2—Recover

Recover Resources (Waste) by Improving Industry Infrastructure

History of Waste

Our planet, in many ways, is now saturated by waste, and we are not talking about bodily waste that comes from humans and wildlife. The waste we are talking about is the byproduct, waste, generated from human activities related to manufacturing, shipping, and consuming products or materials. Some of these materials are food, food packaging, plastics, paper, and wood. Large quantities of material waste from the last hundred years have accumulated quickly. Why so quickly? As discussed in step 1, respect, there are many contributing factors, such as the advancement in technology and the growth in human development and population size. These advancements have helped increase the volume of material waste that has accumulated this last century. Has that waste been efficiently managed? We will now discuss here in step 2, recover, how the United States as a whole and states like Michigan recover waste to the best of their abilities and strive for more sustainable management of waste.

Waste comes in all forms, and is hazardous, nonhazardous, solid, or airborne. The waste that we focus on in this study is solid waste, a type of waste that is typically easier to deal with. The Resource Conservation and Recovery Act (RCRA) states that "solid waste" means any garbage or refuse, sludge from a wastewater treatment plant, water supply treatment plant, or air pollution control facility and other discarded materials resulting from industrial, commercial, mining, and agricultural operations, and from community activities. Nearly everything we do leaves behind some kind of waste. The definition of solid waste is not limited to wastes that are physically solid. Many solid wastes are liquid, semisolid, or contain gaseous material (US Environmental Protection Agency n.d.).

A solid waste is any material that is discarded by being:

> Abandoned: The term abandoned means thrown away. A material is abandoned if it is disposed of, burned, incinerated, or sham recycled.
>
> Inherently Waste-Like: Some materials pose such a threat to human health and the environment that they are always considered solid wastes; these materials are considered to be inherently waste-like. Examples of inherently waste-like materials include certain dioxin-containing wastes.
>
> A Discarded Military Munition: Military munitions are all ammunition products and components produced for or used by the US Department of Defense (DOD) or US Armed Services for national defense and security. Unused or defective munitions are solid wastes when:
>
> - Abandoned (i.e., disposed of, burned, incinerated) or treated prior to disposal;
> - Rendered nonrecyclable or nonusable through deterioration; or

- Declared a waste by an authorized military official. Used (i.e., fired or detonated) munitions may also be solid wastes if collected for storage, recycling, treatment, or disposal.

Recycled in Certain Ways: A material is recycled if it is used or reused (e.g., as an ingredient in a process), reclaimed, or used in certain ways (used in or on the land in a manner constituting disposal, burned for energy recovery, or accumulated speculatively). Specific exclusions to the definition of solid waste are listed in the Code of Federal Regulations (CFR) at 40 CFR section 261.4(a). Many of these exclusions are related to recycling. (US Environmental Protection Agency n.d.)

As addressed previously in the roots of sustainability section, the EPA was created in 1970 to address environmental issues and create policies to protect the natural world. The EPA formed the Clean Air, Clean Water, and Endangered Species Acts right away. However, in 1976, the EPA decided that additional and direct protection measures needed to be enacted to better manage waste. Therefore, the Resource Conservation Recovery Act was formed and enacted in 1976.

RCRA, which amended the Solid Waste Disposal Act (SWDA) of 1965, set national goals for:

- Protecting human health and the environment from the potential hazards of waste disposal.
- Conserving energy and natural resources.
- Reducing the amount of waste generated.
- Ensuring that waste is managed in an environmentally sound manner.

To achieve these goals, RCRA established three distinct, yet interrelated, programs:

- The solid waste program, under RCRA Subtitle D, encourages states to develop comprehensive plans to manage nonhazardous industrial solid waste and municipal solid waste, sets criteria for municipal solid waste landfills and other solid waste disposal facilities, and prohibits the open dumping of solid waste.
- The hazardous waste program, under RCRA Subtitle C, establishes a system for controlling hazardous waste from the time it is generated until its ultimate disposal —in effect, from "cradle to grave."

The underground storage tank (UST) program, under RCRA Subtitle I, regulates underground storage tanks containing hazardous substances and petroleum products. (US Environmental Protection Agency n.d.)

Product and packaging waste make up a great majority of total waste. This type of material waste is classified as MSW, which covers many types of waste. According to the EPA, MSW, more commonly known as trash or garbage, consists of everyday items we use and then throw away, such as plastic or cardboard products and food packaging, grass clippings, furniture, clothing, bottles, food scraps, newspapers, appliances, paints, and batteries. These come from our homes, schools, hospitals, and businesses (US Environmental Protection Agency n.d.).

Before SWDA, and especially the enforcement of the RCRA in 1976, the practice of dealing with and disposing of waste was poor. Sustainability was not a normal mindset at that time, and sustainable waste management was far from being practiced or adopted. Waste was disposed of into rivers or lakes, unmonitored dumpsites, burned, or thrown out somewhere in the environment. The situation was not good, and the United States is still dealing with the ripple effects from bad practices. In many cases, the modern situation regarding waste management is still suboptimal. Globally, uncontrolled dumping of waste is more problematic. Developing nations are working through some of the same growing pains the United States went through in the '60s and '70s. Those growing pains (adoption of sustainable waste management processes) are:

1. Awareness and knowledge of waste, and how, if uncontrolled, it disturbs the health of the environment,
2. Policy implementation and education to the public,
3. Adoption and funding for infrastructure to monitor, measure, collect, dispose or contain, recycle and reduce waste, and
4. Continuous innovation and improvement of sustainable waste management.

The United States was able to learn relatively quickly and to implement a sustainable waste management process to fight waste and its negative impact on the natural world. Maybe the United States can support developing countries in their pursuits to sustainably manage waste. For the sake of Earth, it might be a wise decision. There is still so much room for improvement here on our own soil, and it will be crucial for society to act swiftly before the problem gets out of control, as it has in the past.

Figure 12—Global Population Projections, 2011

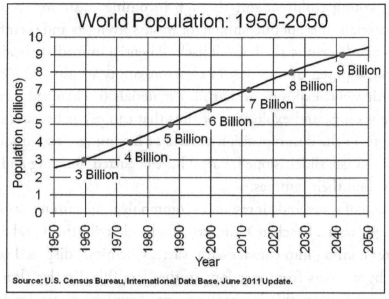

Source: US Census Bureau, International Data Base, June 2011 Update.

Dumping

Before sustainable waste-management practices were adopted (e.g., landfilling, incineration electric plants, composting, or anaerobic digesters and recycling plants), there were dumps. A dumpsite is an area where people pile or toss waste that is discarded because there is no other perceived solution or perceived value for dealing with it. These practices are not sustainable and have mostly been discontinued in the United States; however, as previously mentioned, it still happens in developing nations. The first known type of dump was in Athens, Greece, in 500 BC. Local laws at that time dictated that people dispose of waste at least one mile from the city (Bradbury 2017).

History of dumping:

It is useful to first describe what would generally be considered poor or unsustainable landfill operations; in some locations these might be more commonly described as open or uncontrolled dumps rather than landfills. Historically in developed nations, and currently in many parts of the developing world, waste is disposed of not only in a manner considered unsustainable, but in one that poses risk of direct harm to human health and the environment. The figures below illustrate common conditions at dumps throughout the developing world, and the environmental and human health challenges they present. Economic realities in many nations result in a large human presence at landfill sites, scavengers who are not officially associated with the daily operation of waste disposal. People, often including young children, sort through incoming waste for recovery of salable materials. It is not uncommon for waste scavengers and their families to live on or adjacent to the landfill itself. Potential immediate health risks include those posed by working in close proximity to waste vehicles and heavy equipment, exposure to harmful materials or chemicals, exposure to disease vectors, and explosions or fires that can occur because of gases produced from the decomposition process or incoming reactive wastes. In some cases, waste slides (slope failures) have occurred, burying and killing scavengers and their families.

Pollution of water and air resources commonly results from uncontrolled landfilling of waste. Leachate is the term used to describe the liquid resulting from water coming into contact with waste. Chemicals disposed of in the waste or byproducts from reactions in the [landfill] dissolve (leach) into the water, and when this leachate emerges from the waste [and] enters groundwater or a surface water stream, a risk is posed to those consuming or coming in contact with the affected water resource. Gases and particulate matter can also be released to the environment. Gases produced from the waste decomposition process of food and chemicals merged with other forms of waste, primarily the GHG methane, which posed a potential risk of explosions and fires, and also act as a carrying mechanism for other chemicals in the landfilled waste, many of which may be toxic to humans. Particulates can be released from landfill fires or as dust disturbed as part of landfill operations. Uncontrolled landfilling can pose a threat to ecological resources. Surface water resources contaminated as a result of waste disposal often have reduced dissolved oxygen levels, thus diminishing the ecological health of the water body and potentially resulting in the growth and spread of disease-carrying organisms. Without forethought in appropriate locations for landfills, important ecologic areas are destroyed as a result of waste disposal.

Different Dumpsites with scavengers.
Source: Sustainable Practices for Landfill Design and Operation (Townsend et al. 2015).

A common example is the filling of wetlands as means of reclaiming land. Lastly, indiscriminate disposal of waste through land disposal represents a less than desirable practice from a materials and resource management perspective. Recovery of materials does take place by those sequencing the waste stream, but much more material recovery potential remains buried in the landfill, both in terms of resources and energy. (Townsend et al. 2015, 3–5).

The global situation of waste management is certainly hazardous. There is hope, though, because the United States waste management system has improved, which means others can too. However, the infrastructure is still relatively new, and the waste situation was fairly hazardous not long ago in all of the United States, not just in Michigan. The website zerowasteamerica.org provides an idea of what that situation used to be. According to EPA data, there are around ten thousand old municipal landfills and dumps. Back in the day, every town had its own dump site (and many businesses and factories also had their own). According to the 1997 United States Census, there were 39,044 general-purpose local governments in the United States: 3,043 county governments and 36,001 subcounty general purpose governments (towns and townships). There is a high probability that there are more abandoned commercial, private, and municipal dumps than the ten thousand estimated by the EPA (Zero Waste America n.d.). It is hard to say exactly how much waste is still floating around the planet via land or water, or how much total waste has gone to unmonitored dumpsites. There is no doubt that the health hazards near those dumpsites were severe. Brownfield studies also provide strong examples for how soil and water can be severely contaminated by waste.

Old dumpsite in Grand Rapids, Michigan area.
Sources: City of Grand Rapids Archives and Records Center and Kent County Dept. of Public Works.

Littering

Preceding the dumping mentality, there was general pollution and littering. Unfortunately, some of that still occurs in Michigan, other parts of the United States, and around the globe every day. Pollution is contamination—meaning something that is not natural to the environment—and has a negative impact on the health and general well-being of a particular area. Littering is the act of releasing or disposing of manmade waste, wherever one so chooses. Dumping is semicontrolled, whereas polluting or littering is not controlled at all, making it extremely difficult to measure how detrimental it can be to the environment. Unfortunately, littering and pollution still happen every day where humans live and travel. Some of it is done by accident, some is done intentionally, and some occurs due to a lack of education. If everyone was educated about the impacts of pollution and littering on the environment, would each person stop polluting and littering? Or if waste was more valuable, would people choose not to toss it and instead use its value? These are both important questions and should be taken seriously.

Major examples of littering include:

- Intentional: A person tosses an empty soda can or a plastic bottle from a moving car window on the highway because it's the easiest and most convenient method for disposing of something. The individual has no use for that item anymore.
- Accidental: An empty plastic bag accidently blows off a fishing boat in the Pacific Ocean because of the wind.
- Lack of Awareness or Education: A person tosses scrap wood or cardboard onto his or her property waste pile, not realizing the value of the waste or methods to repurpose it.

Movements, protection agencies, and policies (e.g., Earth Day or the formation of the EPA and the RCRA) have helped control and limit pollution in tremendous ways. In 1990, the EPA formed a new act to raise awareness, educate, and try to reduce pollution to greater extents. This act was called the Pollution Prevention Act.

Pollution Prevention, EPA Statement of Definition:

(Pursuant to the Pollution Prevention Act of 1990 and the Pollution Prevention Strategy)—Under Section 6602(b) of the Pollution Prevention Act of 1990, Congress established a national policy that:

* Pollution should be prevented or reduced at the source whenever feasible;
* Pollution that cannot be prevented should be recycled in an environmentally safe manner whenever feasible;
* Pollution that cannot be prevented or recycled should be treated in an environmentally safe manner whenever feasible; and
* Disposal or other release into the environment should be employed only as a last resort and should be conducted in an environmentally safe manner.

Pollution prevention means "source reduction," as defined under the Pollution Prevention Act, and other practices that reduce or eliminate the creation of pollutants through:

* Increased efficiency in the use of raw materials, energy, water, or other resources, or
* Protection of natural resources by conservation.

The Pollution Prevention Act defines "source reduction" to mean any practice which:

* Reduces the amount of any hazardous substance, pollutant, or contaminant entering any waste stream or otherwise released into the environment (including fugitive emissions) prior to recycling, treatment, or disposal; and
* Reduces the hazards to public health and the environment associated with the release of such substances, pollutants, or contaminants. (Habicht II 1992)

Legal policy and penalties in the form of fines or jail time have been established on a national level to attempt to control and reduce littering. Overall, many states have been successful in reducing pollution and littering and try their best to enforce the policies. It is difficult to monitor every instance of littering. As of 2012, there were approximately 1.4 million law-enforcement officers of a total population of more than three hundred million in the United States (US Department of Justice 2016). So, it is nearly impossible at this time to monitor every single littering act unless new monitoring technologies are adopted. In Michigan, if you are caught littering, you will be charged between eight hundred dollars and five thousand dollars, depending on how many cubic feet of waste were involved (Michigan Legislature 2017).

Litter cleanup per state:

States spend millions of dollars each year to clean up littered roadways, parks, and coastal areas. In addition to the direct cost of litter removal, litter also harms the environment, property values and other economic activity. The most common types of litter are food packaging, bottles, cans, plastic bags, paper and tobacco products. States can discourage littering through a variety of methods, one of which is to create and enforce criminal penalties that punish unwanted behavior. While all states have some type of litter law, penalties vary widely, based on the amount, type, and location of litter. In

10 states, for example, the weight or volume of litter determines the severity of the crime. Other states focus on the type of litter, imposing penalties for dumping large items, such as furniture or major appliances. Many states have also enacted legislation to address littering in certain places, such as public highways, coastal areas, and recreational areas.

For relatively minor cases, courts typically impose a fine and may order litter cleanup or community service. Fines range from $20 in Colorado to $30,000 in Maryland. In more serious cases, offenders may be subject to imprisonment, with sentences ranging from 10 days in Idaho to six years in Tennessee. Laws in Maryland, Massachusetts, and Louisiana also provide for suspension of a [violator's] driver's license in certain cases. Penalties in all states typically increase for subsequent convictions. (National Conference of State Legislatures n.d.)

Street sign on a Michigan country road.
Source: Tyler Kanczuzewski.

There is some good news when it comes to littering. According to a study done in 2009 by Keep America Beautiful (KAB), littering has decreased by roughly 61 percent since 1968. Their findings increased awareness, showing how costly and detrimental littering can be. KAB also estimated that there were roughly fifty-one billion pieces of litter on

US highways, and that is just the highways! Last, the KAB study revealed the estimated cost of littering. According to this 2009 report, litter has high costs for businesses and governments and reduces property values. An estimated $11.5 billion is spent on litter abatement and cleanup activities each year, and the number probably underestimates the true costs (Schultz and Stein 2009, 6). These numbers were from about twelve years ago, so it would be intriguing to see what improvements have been made since 2009. Because the population has grown, business has grown, and human development has grown since 2009, one could imagine that littering and pollution is still occurring in large volumes. Figure 13 shows a breakdown of the types of waste that were typically found according to the 2009 KAB study.

Figure 13—Aggregate Composition of Litter on All US Roadways, 2009

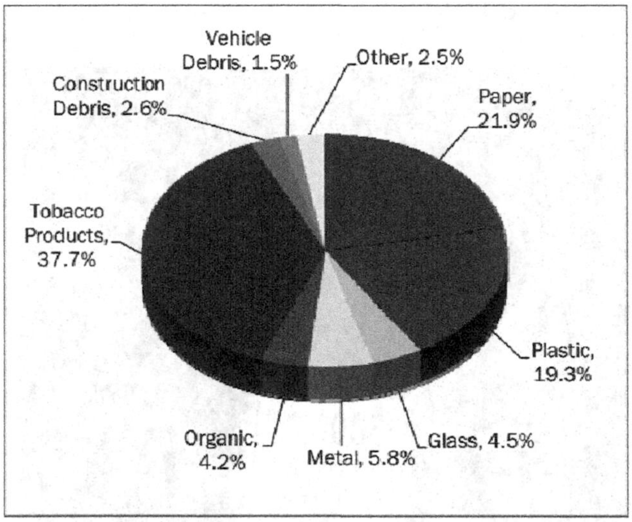

Source: Keep America Beautiful, 2009 Executive Summary of Litter in America by Schultz and Stein.

Recent studies suggest that water pollution—specifically in oceans—is becoming an epidemic. These studies suggest that severely negative consequences are impacting the environment at unprecedented levels. Where did this waste come from, what is it, and how much is now predicted to be in the oceans and in our waterways, like the Great Lakes? Can the damage be undone? We will take some time to dig into what some call the biggest modern epidemic: water and ocean pollution.

Water and Plastics

Water is the key source of all life on the planet. Simply put, without water, we wouldn't be here. Water is imperative, and fortunately, there is a lot of it. As mentioned earlier in this study, water makes up roughly 70 percent of the surface of the Earth, and 97 percent of that is in the oceans. The oceans are made up of mostly salt, so to be potable, they would need to be desalinated. Desalinization is the process of removing the heavy concentrations of salt from the water. Also, as mentioned earlier, the Great Lakes make up roughly 20 percent of the world's usable fresh water, and most of the remaining fresh water is frozen in the North and South Poles. According to the US Geological Survey, there are over 332,519,000 cubic miles of water on the planet. A cubic mile is the volume of a cube measuring one mile on each side. Of this vast volume of water, NOAA's National Geophysical Data Center estimates that 321,003,271 cubic miles is in the oceans. That's enough water to fill about 352,679 quadrillion gallon-sized milk containers (National Ocean Service by NOAA n.d.). This data shows that our planet is blessed with a lot of water, and it would arguably be hard to pollute it all based on its sheer quantity. Unfortunately, research is showing that there is an enormous amount of water pollution happening everywhere because waste is everywhere. Furthermore, research is showing that plastic waste is wreaking havoc and is more detrimental that previously thought to the aquatic ecosystem. Plastic also makes up the brunt of the waste created by products and packaging from human activities.

Seahorse latched on to a plastic cotton swab.
Source: National Geographic, photographer Justin Hofman.

Four Critical Pillars Provided by Water:

- Drinking: 332,519,000 cubic miles of water
    - Water on Earth is plentiful. However, water must be uncontaminated or desalinated to drink. It can be very costly to purify or desalinate water.

- Oxygen and Carbon Sequestration: 70 percent of the world's oxygen and 30 percent of CO_2 sequestration.
  - The oceans provide 70 percent of Earth's total oxygen. The oxygen comes from plants that live in the oceans: phytoplankton, kelp, and algal plankton. And 28 percent of other available oxygen comes from rainforests (National Geographic, n.d.).
  - The ocean holds forty-two times more carbon than the atmosphere and has sequestered 30 percent of recent human induced CO_2 emissions (World Resources Institute 2022).
- Food: Capacity to feed billions of people every day
  - Water, particularly oceans, provides an abundant amount of fish and other types of seafood (Oceana n.d.).
  - The amount of living species in the ocean is unknown.
  - In 2020, global production of fisheries was 178 million tons (The World Bank n.d.).
- Economy: Oceans and waterways contribute more than trillions of dollars annually to the overall economy, according to the Organization for Economic Cooperation and Development (OECD).
  - In 2020, fisheries generated more than $1.5 trillion in economic opportunity. In 2019, aquatic foods provided for over 3.3 billion people and 20 percent of their average animal protein intake (The World Bank n.d.).

These four pillars provide clarity and create awareness about how crucial water is. How is the world actively protecting our precious water? Recent statistics show that civilization is doing a poor job of protecting our water from pollution, littering, and various waste. Can we stop garbage from flowing into or being disposed of into our waters? Time and technological innovations will tell. We will address in step 3, reinvent, how organizations are fighting against and preventing solid waste from entering our waters.

Have you heard of the Great Pacific Garbage Patch (GPGP)? According to a study by the Ocean Cleanup, the GPGP covers an estimated surface area of 1.6 million square kilometers in the Pacific Ocean, an area twice the size of Texas or three times the size of France. It is estimated to hold approximately eighty thousand tons of waste, a weight equivalent to that of five hundred jumbo jets (The Ocean Cleanup n.d.). GPGP is one of five major ocean garbage patches in the world. These garbage patches comprise mostly of plastic, and because most plastic is less dense than water, it floats and forms plastic garbage vortexes, or gyres. Natural ocean currents bring garbage together in these vortexes, and some of the garbage—particularly plastics—have been floating and degrading in these vortexes for more than a decade.

Let's talk plastic. Here is some basic info about the science behind plastics.

Science of Plastics:

Definition: Plastics are a group of materials, either synthetic or naturally occurring, that may be shaped when soft and then hardened to retain the given shape. Plastics are polymers. A polymer is a substance made of many repeating units. The word *polymer* comes from two Greek words: poly, meaning many, and meros, meaning parts or units. A polymer can be thought of as a chain in which each link is the *"mer,"* or *monomer* (single unit). The chain is made by joining, or *polymerizing,* at least 1,000 links together. Polymerization can be demonstrated by making a chain using paper clips or by linking many strips of paper together to form a paper garland.

Examples: Naturally occurring polymers include tar, shellac, tortoiseshell, animal horn, cellulose, amber, and latex from tree sap. Synthetic polymers include polyethylene (used in plastic bags); polystyrene (used to make Styrofoam cups); polypropylene (used for fibers and bottles); polyvinyl chloride (used for food wrap, bottles, and drain pipe); and polytetrafluoroethylene, or Teflon (used for nonstick surfaces). Although many polymers are hydrocarbons that contain only carbon and hydrogen, other polymers may also contain oxygen, chlorine, fluorine, nitrogen, silicon, phosphorus, and sulfur.

Natural polymers such as cellulose and [latex] were first chemically modified in the 19th century to form celluloid and vulcanized rubber. The first totally synthetic polymer, Bakelite, was produced in 1907. The first semisynthetic fiber, rayon, was developed from cellulose in 1911. However, it was not until the global disruption caused by World War II, when natural sources of latex, wool, silk, and other materials became difficult to obtain, that synthetics were mass produced. Synthetic rubber was needed for tires, and nylon was needed as a replacement for silk for parachutes. Today synthetic polymers in the form of plastics are in wide use, and the plastics industry is one of the fastest growing in the United States and around the world. The industry produces approximately 150 kilograms of polymers per person annually in the United States. (Science History Institute n.d.)

The invention of plastic has been revolutionary. Plastic has changed our lifestyles, created big businesses, and, in many ways, has made life better. Plastic can simultaneously be strong, durable, and flexible. On top of that, plastic can last a very long time before degrading or decomposing, in some cases, lasting two hundred to a thousand years or more. Plastic is currently in most of our consumer products. It makes up most of our modern packaging and containerization. For example, most of the food we buy at the local grocery store is sealed and packaged in plastic. Some believe food waste and food packaging comprise about 50 percent of modern landfill waste. Most drinking straws are made of plastic, and there are now many straws in the oceans' garbage patches and on beaches throughout the world. Plastic drinking containers make up a large portion of the packaging pollution that negatively impacts our waters and lands. According to *National Geographic*, last year, the Coca-Cola Company, one of the world's largest producers of plastic bottles, acknowledged for the first time just how many it makes: 128 billion a year. Companies like Nestlé, PepsiCo, and others produce similar amounts of bottles annually (National Geographic 2018). Also, according to the same *National Geographic* study, world production of plastic has increased exponentially, from 2.3 million tons in 1950 to 162 million in 1993 and 448 million by 2015. Now, roughly 40 percent of the 448 million tons of plastic produced every year is disposable, and much of it is used as packaging for food and beverages, which is typically discarded the same day it was purchased (see Figure 14). Production has grown so rapidly that virtually half the plastic ever manufactured has been made in the past fifteen years (National Geographic 2018). The invention of plastic was a significant feat by humans, displaying scientific ingenuity. Some plastics are made up of resources that come from Earth (nonsynthetic), and some come from synthetically created resources. Both forms consume resources (e.g., electricity, fossil fuels, and water) to be engineered and manufactured. As we know, there is not an endless supply of these resources. Are we treating plastic as a scarce resource? If not, should we start doing so? In 2018, the EPA reported that over twenty-seven million tons of plastic were not recycled but went to landfills (U.S. Environmental Protection Agency, 2018). Plastic production has boomed over the last decade, presenting immense opportunities for recyclers and manufacturers to repurpose or recycle various plastics.

Figure 14—Increased Plastic Production and Consumption

Cumulative plastic waste generation and disposal

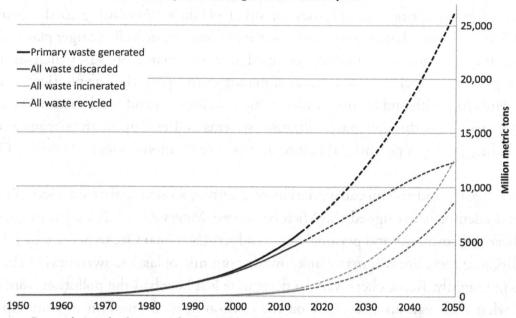

Source: Research Article, Science Advances, Vol. 3 No. 7, Production Use and Fate of all Plastics Made
Link: https://www.science.org/doi/10.1126/sciadv.1700782?cookieSet=1
Authors: Roland Geyer, Jenna R. Jambeck, Kara Lavender Law

According to the Ocean Cleanup project, roughly 1.15 million to 2.41 million tons of plastic enters the oceans each year from rivers (The Ocean Cleanup n.d.). According to a 2018 study done by *National Geographic*, the numbers are even bigger. The study claims that no one knows exactly how much unrecycled plastic waste ends up in oceans. In 2015, Jenna Jambeck, a University of Georgia engineering professor, estimated that between 5.3 million and 14 million tons of plastic waste enters the oceans each year in coastal regions alone. Jambeck and her colleagues believe that most of it isn't thrown off ships, but is dumped carelessly on land and in rivers, mostly in Asia. It's then believed to get blown or washed into the sea (National Geographic 2018). In the Great Lakes Basin, the problem isn't any better. According to the Alliance for the Great Lakes, more than twenty-two million pounds of plastic pollution ends up in the Great Lakes every year. And it never goes away. Instead, it breaks down into smaller and smaller pieces known as "microplastics." Researchers have found alarming amounts of tiny plastic pieces in all five Great Lakes, and these waters provide drinking water for forty million people. They've found microscopic pieces of plastic in drinking water, and even in beer made from Great Lakes water (Alliance for the Great Lakes 2018).

Microplastics is newer terminology. More recent research has pinpointed that plastics in the oceans and the Great Lakes are degrading (usually because of sunlight and natural weather conditions) and are creating microplastic pieces that are as small as five millimeters long. The microplastics are now believed to be entering our drinking water and are potentially being consumed by various aquatic species, as well as birds. It is believed that some fish are mistaking microplastics for small plankton. According to the National Ocean Service, microplastics come from a variety of sources, including larger plastic debris that degrades into smaller and smaller pieces. There are also microbeads, which are a type of microplastic in which very tiny pieces of manufactured polyethylene plastics are added as exfoliants to health and beauty products, such as cleansers and toothpastes. These tiny particles easily pass through water filtration systems and end up in the oceans and the Great Lakes, posing a potential threat to aquatic life (National Ocean Service - Ocean Facts).

Microplastic and microbead pollution was getting so severe that on December 28, 2015, President Obama signed the Microbead-Free Waters Act of 2015, banning plastic microbeads in cosmetics and personal-care products (National Ocean Service by NOAA. n.d.). Because there are still many unknowns about microplastics, awareness of the issue is growing rapidly. Researchers want to determine just how bad the pollution from them is and what their various effects are on the environment. They want to come up with solutions to minimize or reverse the negative impacts on our natural world. There are teams and organizations like the NOAA that are leading marine debris programs to research the issue of microplastics as well as various other potential pollutants from the plastic waste floating in our waters. What can be done to reduce plastic waste, and could plastic waste be reinvented and seen as more valued and reusable?

Colored plastic polymer granules isolated.

Dynamic view of heavy pollution (plastic bottles) on idyllic lake and mountain landscape.

Garbage and plastic waste on an unknown beach.

Landfilling to Manage Waste

The first legitimate method adopted to more sustainably manage waste as a society came in the form of landfilling. Landfilling was on the frontlines, so to speak, for fighting traditional dumping and littering. Landfilling is the process of hauling solid waste to a designated site (landfill), and strategically compacting the waste in specified areas (cells).

Landfilling dramatically changed the way people thought about waste. Was it because all the waste was hauled away and buried so the public didn't see it, like the old saying, "out of sight, out of mind"?

The first true sanitary landfill in the United States was the Fresno Sanitary Landfill (FSL), in 1937. It was designed by Jean Vincenz. According to Historic Fresno, FSL collected 16,500 tons of waste each month, totaling almost eight million cubic yards of waste by the time it closed in 1987 (Historic Fresno, n.d.).

> History of FSL:
>
> Between the opening in 1937 and its close in 1987, the FSL accepted municipal solid waste from the City of Fresno. While the waste stream composition varied over the years as packaging styles and material use changed, the waste likely included materials such as food waste, paper and packaging materials, metal containers, glass, rubber, wood, leather, plastics, and some household cleaning chemicals, pesticides and herbicides, and automobile battery boxes. The landfill was also open to the public for the disposal of tree trimmings and a variety of rubbish. The overall average total waste stream at the FSL consisted of approximately 16,500 tons of waste per month. The total waste quantity in-place is between 4.7 and 8.0 million cubic yards.
>
> Changes in federal law, especially after 1970, placed much higher environmental standards on landfills than those considered in the 1930s and 1940s. Under the Comprehensive Environmental Response, Compensation and Liability Act (CERCLA), which established the Superfund program, landfills were subject to stricter regulation and financial liability. The FSL was first evaluated by the Superfund program as a result of a notification filed by the City of Fresno Solid Waste Management Division on May 27, 1981. The city began the process of closing the landfill in August 1981. The problem of methane gas was first identified in June 1983. On October 4, 1989, the site was placed on the National Priorities List of Superfund Sites." (Historic Fresno, n.d.)

Landfilling and the infrastructure to collect and contain waste became the norm for many city and county governments, as well as private companies, starting in the late 1930s into the 1940s. Waste management became its own booming industry, along with other booming industries of that time. During the mid-1900s, manufacturing growth, the technological revolution, population growth, and growth in human development in general created the perfect storm for waste compilation. One could argue that innovation such as landfilling had to be introduced to do something proactive about the enormous volumes of waste that were accumulating across the country, or people literally would have become buried by their own waste. Hauling and landfilling was a blessing. But has this method of burying and sanitarily storing away waste become too normal, and is there something better we can do with the waste that still accumulates in landfills every day? See Figure 15 to see what a common landfill design looks like.

Old garbage collection truck.
Sources: City of Grand Rapids Archives and Records Center, and Kent County Department of Public Works

Modern garbage truck picking up waste bin.

Figure 15—Overview of Major Components of Modern Engineered Landfills

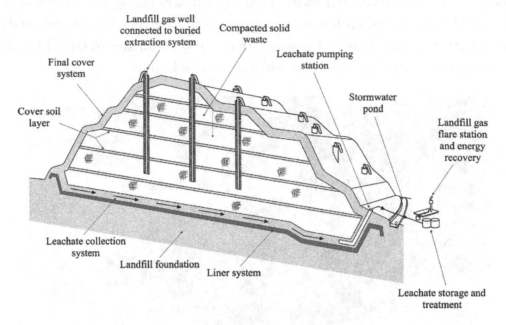

Source: Sustainable Practices for Landfill Design and Operation (Townsend et al. 2015).

Aerial view of a common landfill.

Creation of landfills:

The first step in the evolution of modern landfills from uncontrolled dumps was the development of sanitary landfill practices designed to address immediate human health concerns. The implementation of sanitary landfilling involves several changes to operational practices that focus on minimizing the spread of disease and the occurrence of landfill fires. The placement of waste into defined cells, often constructed in distinct units and compacted in place with heavy equipment, allows more contained and controlled disposal. A critical element in sanitary landfill operation is the routine placement of cover soil on top of recently placed waste to minimize fires, odors and disease vectors. Another key sanitary landfill feature includes site access control, which helps to discourage waste scavenging and properly define the facility's boundary through fencing or similar means. While the evolution of sanitary landfill practices reduced many of the direct human health concerns associated with open dumps, it did not address the two major pollutant emissions associated with landfilled MSW: leachate and landfill gas (LFG). As regulators and scientists began to monitor groundwater quality surrounding landfills, the body of evidence indicating that leachate negatively affected groundwater quality grew (Sawney and Kozloski 1984; Reinhard et al. 1984; Schultz and Kjeldsen 1986). This resulted in many governments requiring MSW landfill construction to include barrier layers for preventing leachate migration out of the landfill and drainage systems allowing the removal of accumulated leachate for treatment before disposal. Many of these technical requirements followed those previously developed for the management of hazardous wastes, a regulatory system designed upon the principle of cradle-to-grave management of wastes that posed an increased risk to human health and the environment. In lined landfill systems, leachate is removed from the landfill and treated prior to its return to the environment. Groundwater surrounding the lined landfill is monitored to assess whether the containment system functions properly. (Townsend et al. 2015)

Before landfilling was adopted, leachate production and gas buildup were accumulating. Therefore, the common landfill was designed and built to capture both pollutants to protect the environment. Leachate can be pumped to a wastewater treatment plant, treated, and then released back into waterways. Gas extraction (mostly in the form of GHG methane) can be used to create electricity or usable biogas. Also, methane and other landfill gases can be treated or burned and then released into the atmosphere at a safer or less harmful level of toxicity to the air. Human engineering can be ingenious and sometimes invisible; however, systems and equipment can fail, leading to a release of leachate into the neighboring groundwater of a landfill site. Untreated gases like methane can leak into the air, leading to a heavy dose of GHG. The greenhouse gas effect is created mostly by human activity and is very harmful to our natural atmospheric system. Heat gets trapped at lower levels and is helping speed up climate change, formerly known as global warming. Landfilling is like a double-edged sword: It's a wonderful solution if it works correctly, but if a failure occurs, it can be devastating to the surrounding natural world (Center for Health, Environment & Justice 2015).

Waste collection at landfill, and bulldozer compacting the waste.
Source: Sustainable Practices for Landfill Design and Operation (Townsend et al. 2015).

Dirt-spreading on landfill.
Source: Sustainable Practices for Landfill Design and Operation (Townsend et al. 2015).

As of 2018, there were more than two thousand active landfills across the country (see Figure 16). According to the EPA Landfill Methane Outreach Program (LMOP), there are 2,455 facilities in operation or listed as candidates for new landfills (see Figure 17). The average size of each landfill is about 130 acres (used land plus available land), and each landfill holds about 5.3 million tons of waste (over 11.2 billion tons of total waste);

however, landfill sizes vary greatly. For instance, the biggest landfill in the country—which is now capped and closed—holds 142 million tons of waste and is 590 acres (US Environmental Protection Agency n.d.). That landfill is the Puente Hills Landfill in Los Angeles. Los Angeles is also one of the largest and most densely populated cities in the United States. According to the EPA, there were about eight thousand active landfills across the country in the late 1980s. At that time, some of the landfills were less regulated and some were even considered dumps. The good news is that the total number of landfills has decreased dramatically; however, that fact does not mean less waste is being generated. Landfills are becoming more efficiently designed, allowing for them to be bigger and to contain more waste. Landfill life spans also vary greatly, based on the capacity of the land or the efficiency of the design. Once a landfill is capped and closed safely, it is believed that, after thirty years, it can be considered safe for potential reopening. Would this mean that companies could then mine an old landfill for waste that has completely decomposed?

Of the total 2,455 landfills that are active or are potential new candidates, taking the average size of 130 acres each results in a total of more than 320,000 acres of land used and reserved for landfills. Since the United States has about 2.3 billion acres of land, used-and-reserved land equates to less than 1 percent. The total acres of land needed for landfilling is relatively small right now; however, if businesses, the population, and waste accumulation grows, the need for new landfills will increase. Hopefully, newer strategies can be implemented to reduce waste, recycle, and repurpose more and more materials. Also, can more products be made to decompose faster that are better for environmental health?

Figure 16—Total Open and Active Landfills in the
United States (Two Thousand-Plus)

Sources: SaveOnEnergy, Land of Waste Study, and EPA.GOV, 2015.
Link: https://www.saveonenergy.com/land-of-waste/.

Figure 17—Project Landfill Data by State, Plus Candidates (2,400-Plus)

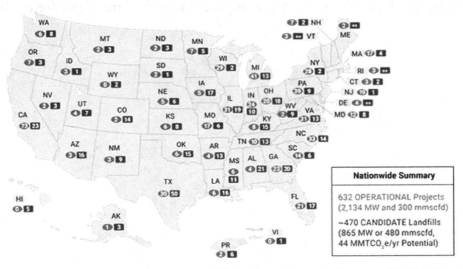

Source: US Environmental Protection Agency, Landfill Methane Outreach Program (LMOP Database), 2018.

As of 2018, the total waste generated by the United States hit an all-time high at 292.4 million tons of MSW, about 4.9 pounds per person per day (U.S. Environmental Protection Agency, 2018). More than 50 percent of that waste went to landfills. Total revenue generated by the waste industry was well in the billions of dollars. According to the EPA, in 2015, the average price charged per ton to dump waste at the common landfill was around US$48. The infrastructure for waste management and landfilling varies by region and by government. It is fairly common that county and city governments (and departments of public works) operate their own garbage retrieval systems as well as their own landfills. However, in many cases, both private and public companies will operate waste-management services, such as garbage pickup, landfilling, and recycling. Waste Management is one major company that conducts business in the waste industry and is publicly traded. Republic Services and Waste Connections are two other major waste-management firms, and they are also publicly traded.

There are many benefits to landfilling, ranging from asethics to economic purposes. In many cases, landfilling makes sustainable sense, and it seems the optimal solution for dealing with waste. In some regions, for instance, Kent County, Michigan, space is limited for collecting new waste. But what other sustainable options existin areas like Kent County? Is landfilling potentially cheap, and should waste be valued higher? Is the economic generation created by the waste industry now more important than the health of the environment? Is the methane gas generated by landfills and converted to electricity an important part of the United States' energy future?

According to the EPA, methane is a potent GHG, and it is twenty-eight to thirty-six times more effective than CO_2 (carbon dioxide, another potent GHG) at trapping heat in the atmosphere over a hundred-year period (US Environmental Protection Agency n.d.). If not controlled properly, methane and leachate can severely harm our natural world. What are we willing to put at stake to better protect our environment and be good stewards of our planet? What measures can be put in place to improve sustainable waste management? See Figure 18, which shows the number of tons of trash produced per person by state over the course of a year. Remember, as of 2018, the average person throws out 4.9 pounds of waste each day.

Figure 18—Tons of Trash in Landfills Per Person by State, 2015

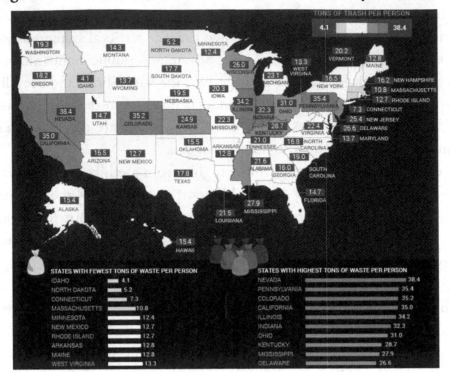

Sources: SaveOnEnergy, Land of Waste Study, and EPA.GOV, 2015.
Link: https://www.saveonenergy.com/land-of-waste/.

Each geographic region and state produce different amounts of waste. Populations and economic systems vary. Some states have more resources or more economic opportunities. For the most part, regions and states that have higher populations and more industries produce more waste and more landfill gas. See Figure 19, which shows estimated landfill gases (methane per cubic feet of gas each day, MMSCFD) produced by state.

Figure 19—Landfill Gas Produced by State, 2015

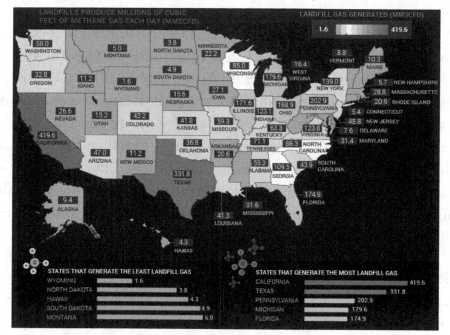

Sources: SaveOnEnergy, Land of Waste Study, and EPA.GOV, 2015.
Link: https://www.saveonenergy.com/land-of-waste/.

Facts on landfill methane, and electricity generation:

Landfills are the third largest source of anthropogenic methane in the United States. According to the US Environmental Protection Agency (EPA), landfill gas (LFG) comprises 17.7 percent of all US methane emissions. Landfill methane in 2011 accounted for 103 million metric tonnes of carbon equivalent released into the atmosphere. Methane is a short-lived climate pollutant with significant warming potential, and over a twenty-year period, one ton of methane causes seventy-two times more warming than one ton of carbon dioxide (CO_2). Consequently, the mitigation of methane from existing landfills provides important climate benefits.

Mitigation of LFG can provide health benefits as well. Landfill gas is comprised of approximately 50 percent methane and 50 percent (CO_2), with trace levels of other compounds, including nitrogen, oxygen, hydrogen, and nonmethane organic compounds (nmocs) such as ammonia and sulfides. Nmocs include hazardous air pollutants that can increase the risk of cancer, cause respiratory issues, and produce strong and unpleasant odors. To mitigate both health and environmental impacts, the EPA currently regulates LFG from very large municipal solid waste (MSW) landfills, which must capture and safely dispose of methane and nmocs from LFG. This process is typically accomplished either by flaring the gas or by converting the gas into energy.

    <u>Methane to Energy:</u> To encourage landfill operators and development partners to capture and harness LFG, the EPA created the Landfill Methane Outreach Program (LMOP) in 1994. As of October 2012, there are 605 operational energy projects in 48 states, and LMOP estimates that another 400 additional landfills are good candidates for energy projects. Together, the operational landfills produce approximately 15 billion kilowatt-hours (kwh) of electricity and 100 billion cubic feet of LFG for direct use annually. In 2012 alone, the amount of methane removed was equivalent to eliminating the (CO_2) emissions from approximately 240 million barrels of oil consumed. (Cross, John-Michael 2013)

Electricity generation of fifteen billion kilowatt-hours (kwh) and one hundred billion cubic feet of LFG for direct use sounds impressive. Converting methane to energy (electricity) is better than it being wasted. However, do the positives of energy production outweigh the negatives, such as potential pollution?

Incineration and Waste-to-Energy

An alternative to the ridding of waste came in the forms of incineration and waste-to-energy (WTE) plants. The major difference between the two is that old incineration facilities used to burn trash to get rid of it; whereas modern WTE plants burn trash to generate electricity that can be used in our electrical grid. Incineration and WTE plants can be optimal, especially if land is scarce or alternative forms of electrical generation are needed.

According to the EPA, the first incinerator built in the United States was in 1885 on Governors Island in New York. By the mid-twentieth century, hundreds of incinerators were in operation in the United States. Until the 1960s, little was known about the environmental impacts on water discharges and air emissions from those incinerators. When the Clean Air Act (CAA) came into effect in 1970, existing incineration facilities faced new standards that banned the uncontrolled burning of MSW and placed restrictions on particulate emissions. The facilities that did not install the technology needed to meet the CAA requirements closed. Combustion of MSW grew in the 1980s. By the early 1990s, the United States combusted more than 15 percent of all MSW. Most nonhazardous waste incinerators were recovering energy by this time and had installed pollution control equipment. With the newly recognized threats posed by mercury and dioxin emissions, the EPA enacted the Maximum Achievable Control Technology regulations in the 1990s. As a result, most existing facilities had to be retrofitted with air pollution control systems or be shut down (US Environmental Protection Agency n.d.).

According to the US Energy Information Administration (EIA), municipal solid waste is usually burned at special WTE plants that use the heat from fire to make steam, which is used for generating electricity or to heat buildings. In 2016, seventy-one US power plants generated about fourteen billion kwh of electricity from burning about thirty million tons of combustible MSW. Producing electricity is one major reason to burn MSW. Burning waste reduces the amount of material that would otherwise be buried in landfills, and burning MSW reduces the volume of waste by about 87 percent (US Energy Information Administration 2018).

Grand Rapids, Michigan, is one city that has a WTE plant. Grand Rapids is the second-largest city in Michigan in terms of population. The economy in this city is thriving. Space is somewhat limited in the region, and the critical landfills are reaching capacity. Therefore, disposing and burning trash to create electricity seemed like a great solution. The Kent County Facility, operating as Covanta Kent, Inc., began commercial operation in January 1990. The 9.86-acre facility in Grand Rapids processes 625 tons of municipal solid waste per day, generating up to eighteen megawatts of electricity. Under Covanta's operating contract, the company is responsible for maintaining the facility and fossil-fuel steam plant. Waste is delivered to the facility from Grand Rapids and five surrounding cities.

According to EPA archives, it is estimated that greenhouse gas emissions from US MSW combustion facilities, or WTE plants, range from ten million to twenty million metric tons. The numbers vary depending on the different methods used to estimate the biogenic fraction of MSW. It is a small fraction of the nearly six billion tons emitted by the combustion of fossil fuels. Per unit of electricity produced, the MSW combustion facilities generate fewer GHGs than coal or oil but are comparable to natural gas (US Environmental Protection Agency n.d.). See Figure 20.

Waste-to-Energy Plant Example.

Figure 20—CO$_2$ Emission by Combustion of Fossil Fuels

Fuel	CO2 (pounds per megawatt hour)
MSW	1016
Coal	2249
Oil	1672
Natural Gas	1135

Source: US Environmental Protection.
Link: Agencyhttps://archive.epa.gov/epawaste/nonhaz/municipal/web/html/airem.html.

How sustainable are WTE plants? They provide proof that humans can be innovative to solve problems. But do they truly solve our biggest problem? They do get rid of waste and provide electricity, but does that make them a sustainable solution? In geographic areas where space is limited, is burning waste the best method? How can we best address large volumes? What are the optimal methods for reducing, repurposing, and using waste? More on this topic will be discussed in step 3, reinventing waste.

Recycling and Reuse Evolution

The idea of recycling and reusing materials has existed for some time. Why not reuse or refurbish something instead of spending additional time and resources to rebuild the same thing? The idea is basic, and since we live in a world of scarcity, it makes sense to reuse things, save time, and consume less of Earth's resources … right? But as we talked about in step 1, respect, humanity's pursuit of maximizing happiness and the good of people has generated amazing technological advancements and revolutions (industrial, technological, and digital). A person can now travel the Earth faster and cheaper, indulge in more of the available resources, and start revolutionary businesses that supply a new good or service for humanity to enjoy. Because of these advancements, humans can consume more resources and live longer. And don't forget the significant logistical hurdles that have been overcome through our major revolutions that have changed the way humans consume goods and services and produce waste—improved resource extraction, mass production, cost and convenience, and delivery.

Overcoming logistical barriers has spurred growth in capitalism, business competition, and consumerism. It no longer takes an extensive amount of time to build something new, and our new technologies make it seem like resources are abundant. This movement has given birth to our modern-day globalized economic engine, an engine that provides seemingly unlimited resources that are available at the click of a button. But we don't live in a world of unlimited resources. And waste has been accumulating year over year.

We talked earlier about the roots of sustainability. Resource extraction, waste, and pollution were rapidly growing simultaneously, and something had to be done to help the plant heal. Earth Day originated in 1970 and helped create the EPA, which laid the foundation for environmental protection and sustainable practices. In 1976, the EPA passed the Resource Conservation and Recovery Act to further sustainability efforts related to waste and to encourage resource conservation. Somewhere during this period, a new idea came about; rather, a new way to live. The idea was to "reduce, reuse, and recycle" (the three Rs). The three Rs have now raised great awareness across the country and the world, year after year, and have created a mindset of reducing consumption of products and consumer goods whenever possible, reusing things whenever possible, and to recycle as a last resort. Recycling and the three Rs campaign created wonderful awareness and has been changing people's mindsets about how to deal with waste.

According to TerraCycle, recycling is the process of recovering material from waste and turning it into new products (Terracycle n.d.). Recycling can also mean repurposing an item, such as using it for something else. For instance, a plastic bottle that was originally used for drinking water can be cut in half and reused to scoop dirt in a garden. To think about recycling as a process, think about Earth's natural water cycle (see Figure 21). The cycle, or process, is natural, and water is recycled over and over with virtually no waste. The goal of recycling is to minimize waste or to reuse it. If you can create a circular flow—where something is used and then reused, giving life back to a system or cycle—that would seem to be sustainable. It would make sense that a cycle that gives life back to itself would last forever, and lasting forever would be the highest level of sustainability.

Figure 21—The Water Cycle

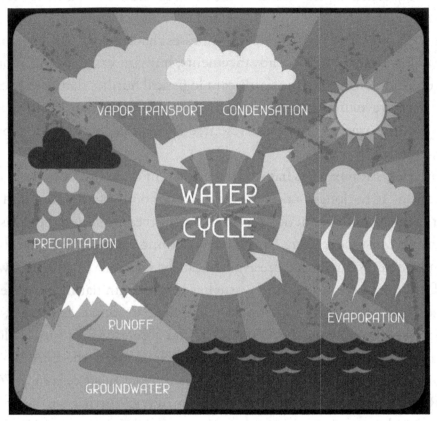

Source: Clip Art Vector, the Water Cycle.

The first known legitimate recycling effort on a broad scale was in Japan around the year AD 1000. The country started a system of reusing waste paper. All of the documents and paper were recycled and repulped into new paper and then sold at local shops across the country (Bradbury 2017). Then, in 1690, recycling began in the United States, when the Rittenhouse Mill in Philadelphia began recycling linen and cotton rags. The paper produced from these materials was sold to printers for use in bibles and newspapers (American Disposal Services n.d.). In 1865, the Salvation Army was founded in London and began collecting, sorting, and recycling unwanted goods. The Household Salvage Brigades employed the unskilled poor to recover discarded materials. The organization and its program migrated to the United States in the 1890s (Bradbury 2017). In 1897, New York City created a materials recovery facility where trash was sorted at picking yards and separated into various grades of paper, metals, and carpet. Burlap bags, twine, rubber, and even horsehair were also sorted for recycling and reuse (Bradbury 2017). In the early 1900s, a campaign was started for developing cities called "waste as wealth" to encourage people to sort their waste and find ways to create value (Jefferson Recycling 2017). The year 1904 was big for aluminum-can recycling, as two plants opened: one in Chicago and one in Cleveland (Bradbury 2017). World Wars I and II (around 1915 and 1940, respectively) created shortages of various resources caused by the war efforts and the stock market fallout. Consequently, several raw materials, such as nylon, rubber, and metals, were rationed and recycled. Many people reclaimed metals and rags to make ends meet (Jefferson Recycling 2017).

A setback to the strengthening recycling industry came in the mid-1950s, climaxing when *Life* magazine published an issue in 1955 called "Throwaway Living." At that time, production and consumption of single-use, disposable plastics was skyrocketing. Advances in technology and science allowed for it to happen, and the products were very convenient (National Geographic 2018). That time period helped create a growth spurt in waste accumulation. A turning point took shape in the 1960s, when the first curbside collections of yard waste, metals, and paper started popping up around the country. Separate waste streams collected at the curb became common as well (American Disposal Services n.d.). Around 1970, the Mobius loop (the first recycling emblem) was designed by Gary Anderson, a twenty-three-year-old college student. The design was the winning entry for an art contest sponsored by a Chicago-based recycled paperboard company to raise environmental awareness among high schools and colleges across the country (Green Dining Alliance, n.d.). The 1970s were extremely pivotal, with the Earth Day movement, the formation of the Environmental Protection Agency (EPA), and the Resource Conservation and Recovery Act (RCRA). Also in the 1970s, Oregon introduced a refund credit on bottles, the first recycling mill was built in Conshohocken, Pennsylvania, and the first curbside recycling-bin program was adopted in University City, Missouri, called "The Tree Saver" (Bradbury 2017).

The 1980s created more excitement and changed recycling efforts even further. In 1981, Woodbury, New Jersey became the first city in the United States to mandate recycling. In 1987, The *Mobro 4000* (nicknamed "the garbage barge") spent months on the ocean, searching for a location to dispose of its garbage cargo to another city along the coast, as New York City had no more room for trash. The saga was widely covered in the media because the barge couldn't lock in a contact to dispose of the waste and has been credited with awakening Americans about solid waste and the importance of recycling (American Disposal Services n.d.). By 1988, the amount of recycling and curbside programs grew to around 1,050. From the late 1980s to the mid-1990s, major strides took place, with the amount of curbside recycling programs growing to more than four thousand, and total recycling centers in the United States passing ten thousand (Bradbury 2017). In 1996, the EPA reported that recycling efforts saved 25 percent of waste from going to the landfill or elsewhere and set a goal of hitting a 35-percent recycling rate. In 2000, the EPA created more awareness and delivered startling news when research showed that waste accumulation in the United States was contributing to climate change, greenhouse gas emissions, and potential damage to the health of the natural world (Bradbury 2017).

Mobius loop, recycling emblem designed by Gary Anderson.

United States (Plus International) Recycling History and Critical Events:

- 1000: Japan—known as a culture to have recycled and resold wasted paper.
- 1690: Philadelphia—the Rittenhouse Mill recycled and resold linen and cotton rags.
- 1865: London—the Salvation Army collected, recovered and recycled unwanted goods.
- 1897: New York City—materials recovery facility divided trash from recyclable materials for resale or reuse.
- Early 1900s: United States—"Waste as Wealth" campaign, movement and mindset change.
- 1904: Chicago and Cleveland—aluminum can recycling for reuse and resale.
- 1910 to 1950: United States—wartime created shortages of certain materials, which boosted reclamation and recycling opportunities.
- 1955: United States—*Life* magazine celebrates "Throwaway Living." Contributes to a new national mindset that single-use plastics are convenient. Waste volume quickly grows.
- 1960s: United States—national adoption of separating waste and curbside collection. Sustainable practices start.
- Late 1960s: Chicago—creation of Mobius recycling emblem, which raises awareness for widespread recycling efforts.
- 1970: United States—Earth Day movement and creation of EPA, an increase in resource protection, conservation and recycling efforts occurs nationwide.
- 1971: Oregon—refund credits (nickel) on recycled bottles comes about and incentivizes people to recycle.
- 1972: Conshohocken, Pennsylvania—first recycling mill built to recover and recycle waste streams.
- 1974: University City, Missouri—first-known curbside recycling service by a city begins. It then incentivizes people and changes mindsets.
- 1976: United States—Passing of RCRA occurs, resource conservation and recovery of waste and materials for reuse.
- 1981: Woodbury, New Jersey—recycling mandated. This forces locals to recycle and even changes lifestyles.
- 1987: United States—*Mobro 4000* trash barge tries to find location to dump waste from New York City. The event raises awareness, showing the world how waste is piling up quickly and how landfills are quickly overflowing.
- 1988: United States—Curbside recycling efforts reaches 1,050 programs. This momentum creates more awareness for fast-growing recycling efforts.

- 1995: United States—Curbside recycling programs reach four thousand, and more than ten thousand recycling centers are listed nationwide. Recycling finally reaches more sustainable levels nationwide.
- 1996: United States—EPA reports 25 percent of waste is recycled and diverted from landfills.
- 2000: United States—EPA admits waste is contributing to the greenhouse gas (GHG) effect and contributing to climate change.
- 2018: United States—EPA reports 23.6 percent (sixty-nine million tons) of MSW is recycled from 292.4 tons of total waste. It's a US record for most trash produced in one year.

Mobro 4000 garbage barge floats near New York City.
Source: Newsday, Ken Sawchuk.

It's safe to say that recycling efforts are part of our way of life now and at the forefront for decision-making and dealing with waste accumulation. Industries, manufacturers, businesses, and the regular person now all look at waste differently. Did major campaigns like the three Rs and "Waste as Wealth" help create the mindset change, or are people finally seeing the value in recovering, reusing, repurposing, and selling waste? Are companies, organizations, and regular homeowners looking at their balance sheets and able to pinpoint more savings or more earnings by either reducing consumption or by repurposing and selling products? All could be part of the change and are playing a role in current sustainable waste management. However, are we still doing enough, and can the United States take sustainable waste management systems to the next level of efficiency and effectiveness? Can we create a new campaign and movement that will create an even bigger mindset change than ever before?

Modern method of assorted recycling bins (above).

In 2017, the United States recycled around 26 percent, or 67 of 268 tons of total municipal solid waste monitored (US Environmental Protection Agency 2018). But in 2018, the percentage of recycled materials decreased to 23.6, or 69 of 292.4 tons. There are currently roughly ten thousand recycling centers nationwide, and over four thousand curbside recycling pickup services. There are also many more forms of recycling and material repurposing facilities across the country. Some of these concepts and business platforms have been around for many years, and some are new. Some are actually very profitable, but others make just enough money to cover basic, overhead expenses. All of these recycling concepts serve a purpose and realize the value in recycling and repurposing items. For some, intent is not just about profits but rather about care for people and the planet. If you counted all these additional recycling and product-repurposing facilities listed next, it would add thousands more recycling-related centers and divert millions of additional tons of potential material waste from going to the landfill.

- Metal scrap yards and facilities
- General scrap facilities
- Electronic recycling centers
- Junkyards
- Used-car resellers
- Surplus goods inventory centers (selling old inventory)
- Thrift stores (e.g., Salvation Army, Goodwill)
- Homeless shelters or ministries in need
- Antique malls
- Reclaimed building-material resellers (e.g., Habitat for Humanity)
- Used books resellers (e.g., Better World Books)
- Blemished food resellers and food pantries (e.g., Imperfect Produce)
- Food or organic matter composting facilities (e.g., from homes or businesses)
- Various used goods and product resellers

Example of recycling and scrap center.

Many of these facilities and concepts are making huge impacts on material waste reduction, which also makes a positive impact on the environment. Some additional examples of material or potential waste reuse include food waste given to livestock on a farm, road construction companies grinding old concrete to repurpose for newer roads, and various used and raw materials fermented to create biomass or biofuel for another purpose. Different forms of recycling now happen in most industries, governments, schools, municipalities, and homes. The recycling movement is necessary, imperative, and can be efficient if managed properly. Modern-day companies are looking seriously into triple-bottom-line efforts (people, planet, profit), and taking part in sustainability initiatives. Some people are calling the modern era of changed business mindset a "green movement." Many modern-day companies are intensively trying to reduce waste and reuse scrap. However, are these companies accurately evaluating the true value of these materials and for the right reasons? Is it cheaper to make or produce something new or from virgin feedstock (raw material used in manufacturing), versus recycling something? We will discuss this further in step 3, reinvent.

Kent County, Michigan, has an established system for waste management, including multiple landfills, recycling drop-off locations and centers, a WTE plant, and an education center. All these locations and services are operated by the Department of Public Works. Kent County's recycling and education center serves as the primary materials recovery facility for residential recyclables generated in homes throughout West Michigan. A materials recovery facility (MRF) is a processing facility where recycling from curbside bins and drop-off recycling centers (that are generally paid for by the public) is sorted and prepared to be sold to processors. The recycling and education center works with about fourteen different processors, and the recycled waste is sold to them by the pound or, in some cases, given for free. Nearly fifty-eight-million pounds of recyclables were delivered to the facility in 2017. These materials were put back into the local economy as feedstock for new products like cereal boxes, glass bottles, toys, packaging, clothing, park benches, and more (Kent County Department of Public Works 2018).

Baled recycled plastics and various waste at a reclamation center.

Composting

Composting is a process that occurs organically in nature; it is the decomposing of plant matter and other earthly materials. By combining insects, invertebrates, bacteria, and organisms with plant matter, you can create rich soil. With that rich soil, one can support flora and fauna and spur new growth. One can mimic nature by composting his or her own waste at home or at his or her business. Composting dates back thousands of years, and evidence shows that it existed in many different civilizations throughout time. The standard operation was (and still is) to take some form of reclaimed material and use it for agriculture. According to *National Geographic* and Dennis Pogue, a historical archaeologist, George Washington was one of the first American composters and helped the concept go mainstream. Washington was even quoted as saying, "For the United States to succeed, we must become better farmers" (National Geographic 2016).

Composting is beneficial on many levels, but a transition from small- to large-scale, and growing infrastructure capabilities, have greatly increased the annual volume of total composted municipal solid waste since the late 1980s. Some cities have started collecting curbside compost, and private companies have been innovating and creating composting systems and solutions for the last few decades. In 2018, the EPA reported the second-biggest year on record, at 24.9 million tons of waste composted, compared with just four million tons reported in 1990. However, roughly thirty-five million tons of food waste went to landfills, making it the single-largest waste stream (US Environmental Protection Agency 2018). Of that thirty-five million tons, a sizable percentage could be used for composting: for example, fruits, vegetables, greens, egg shells, and coffee grounds. The third-largest waste stream reported by the EPA is paper and cardboard, at about eighteen million tons. Fortunately, paper and cardboard are also optimal for composting in a commercial or self-made composting mound, bin, heap, or system.

When food waste enters a landfill, it creates an anaerobic environment, creating methane gas, which is much worse and more harmful than CO_2 (the main known culprit for climate change). Landfills can either burn off the methane or use it to create electricity or biogas. Instead, food scraps and paper products could be delivered to composting systems to create nutrient-rich soil that cultivates new growth and life. Research proves that composting can, in fact, store carbon in the ground, in soil, and through the cycle of creating rich soil and spurring new vegetation. Improved composting at municipalities, commercial institutions, and residences has the potential to greatly reduce landfill waste and capture carbon dioxide (CO_2).

Source: Jon Lowenstein, Noor/Redux, National Geographic.

Biogas Recovery, Anaerobic Digesters and Bioenergy Potential

Aside from composting, there are now innovative systems for biogas recovery, anaerobic digestion, and bioenergy. Figure 22 shows the anaerobic digestion process, a modern-day system and innovation that can convert food waste and organic waste into different forms of energy. According to the American Biogas Council, anaerobic digestion is a series of biological processes in which microorganisms break down biodegradable materials in the absence of oxygen. One of the end products is biogas, which is combusted to generate electricity and heat, or that can be processed into renewable natural gas and transportation fuels (American Biogas Council 2020).

According to the Environmental and Energy Study Institute, the United States produces more than seventy-five million tons of organic waste, though not all is considered in the municipal solid waste category, and in 2018, over thirty-five million tons of food waste was landfilled. Not only can we use waste to create energy, we can minimize pollution and GHG emissions because biogas provides clean, safe, reliable, and renewable sources of baseload energy production in place of coal or natural gas (Environmental and Energy Study Institute 2017).

> Similar to natural gas, biogas can also be used as a source of peak power that can be rapidly ramped up. Using stored biogas limits the amount of methane released into the atmosphere and reduces dependence on fossil fuels. The reduction of methane emissions derived from tapping all the potential biogas in the United States would be equal to the annual emissions of 800,000 to 11 million passenger vehicles. Based on a waste-to-wheels assessment, compressed natural gas derived from biogas reduces greenhouse gas emissions by up to 91 percent relative to petroleum gasoline. In addition to climate benefits, anaerobic digestion can lower costs associated with waste remediation as well as benefit local economies. Building the 13,500 potential biogas systems in the United States could add over 335,000 temporary construction jobs and 23,000 permanent jobs. Anaerobic digestion also reduces odors, pathogens, and the risk of water pollution from livestock waste. Digestate, the material remaining after the digestion process, can be used or sold as fertilizer, reducing the need for chemical fertilizers. Digestate also can provide additional revenue when sold as livestock bedding or soil amendments. (Environmental and Energy Study Institute 2017)

Figure 22—Anaerobic Digestion Process

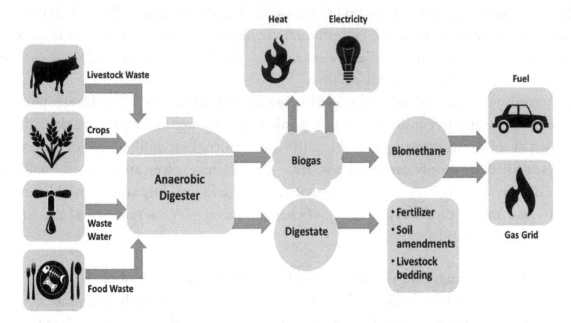

Source: Environmental and Energy Study Institute (EESI), and Sara Tanigawa.

The Complete (MSW) Waste Stream According to the EPA

This section lays out the big picture for the entire MSW stream of the United States, excluding construction and demolition debris. This information is pulled directly from the EPA, as this agency best monitors the entire waste stream of the United States and has been doing so for over thirty-five years. The data below shows material streams and the total waste that is accumulating, as well as where the waste is directed or how it is managed. Construction and demolition (C&D) debris is excluded, but accounts for 600 million tons of total material collected, and 144 million tons landfilled alongside 146 million tons of MSW. That means that in 2018, 290 million tons of material was landfilled all-together.

The US Environmental Protection Agency (EPA) has collected and reported data on the generation and disposition of municipal solid waste (MSW) in the United States for more than 35 years. This information is used to measure the success of materials management programs across the country and to characterize the national waste stream. These facts and figures are based on the most recent information, which is from calendar year 2018. In 2018, in the United States, approximately 292 million tons (United States short tons unless specified) of MSW were generated. Of the MSW generated, approximately 69 million tons were recycled and 25 million tons were composted. Together, about 94 million tons were recycled or composted, equivalent to a 32.1 percent recycling and composting rate. In addition, about 18 million tons of food (6.1 percent) were processed through other food management pathways. More than 34 million tons of MSW (11.8 percent) were combusted with energy recovery. Finally, more than 146.2 million tons (50.0 percent) were landfilled. Information about waste generation and management is an important foundation for managing materials. EPA's Sustainable Materials Management (SMM) approach refers to the use and reuse of materials in the most productive and sustainable way across their entire lifecycle. Through SMM, EPA helps to meet the material needs of the future by providing methods to decrease environmental impacts of materials use while increasing economic competitiveness. This report analyzes MSW trends in generation and management, materials and products, and economic indicators affecting MSW. It also includes a section on the generation and management of construction and demolition (C&D) debris, which is not a part of MSW, but comprises a significant portion of the nonhazardous solid waste stream. (US Environmental Protection Agency 2018)

Official EPA Logo.

Figure 23—Total MSW Generation (1960–2018)

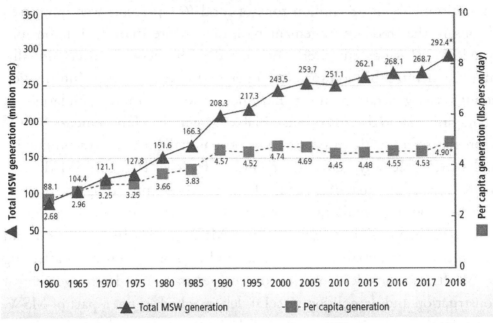

Source: Advancing Sustainable Materials Management: 2018 Facts and Figures (2020).
Author: US Environmental Protection Agency.

In 2018, 292.4 million tons of MWS waste was generated, which was the highest amount ever recorded in one year (see Figure 23). However, 2018 was a somewhat efficient year in terms of the percentage of total waste recycled and composted, at 93.9 percent, as shown in Figure 24.

Figure 24—Total MSW Recycled and Composted (1960–2018)

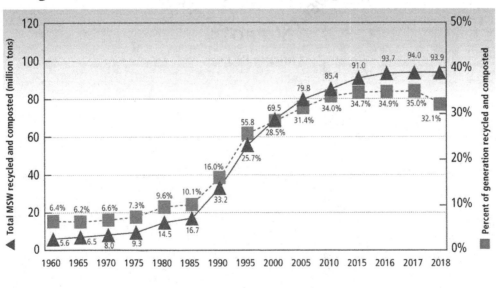

Source: Advancing Sustainable Materials Management: 2018 Facts and Figures (2020).
Author: US Environmental Protection Agency.

Figure 25—Total MSW Percentage Breakdown in 2018

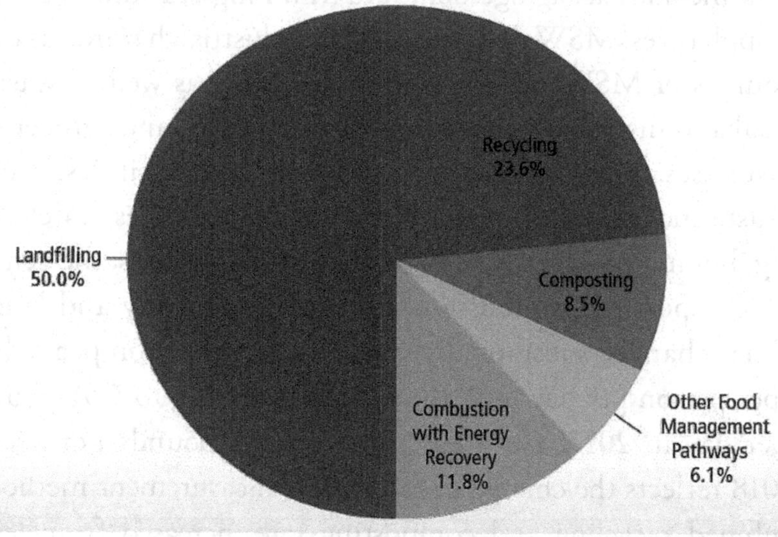

Source: Advancing Sustainable Materials Management: 2018 Facts and Figures (2020).
Author: US Environmental Protection Agency.

In 2018, the United States landfilled 35.28 million tons of food waste: 27.03 of plastics, 17.22 of paper and cardboard, 12.15 of wood, 11.30 of textiles, and 10.53 million tons of steel. Overall, there were 97.1 million tons of material in products that were landfilled. All these materials could be valuable resources and used, rather than dumped and landfilled.

Figure 26—Waste Generation in Millions by Material Type in 2018

Material	Weight Generated	Weight Recycled	Weight Composted	Weight Other Food Management Pathways¥	Weight Combusted with Energy Recovery	Weight Landfilled	Recycling as Percent of Generation	Composting as Percent of Generation	Other Food Management Pathways as Percent of Generation	Combustion as Percent of Generation	Landfilling as Percent of Generation
Paper and paperboard	67.39	45.97	-	-	4.20	17.22	68.2%	-	-	6.2%	25.6%
Glass	12.25	3.06	-	-	1.64	7.55	25.0%	-	-	13.4%	61.6%
Metals											
Steel	19.20	6.36	-	-	2.31	10.53	33.1%	-	-	12.0%	54.9%
Aluminum	3.89	0.67	-	-	0.56	2.66	17.2%	-	-	14.4%	68.4%
Other nonferrous metals†	2.51	1.69	-	-	0.08	0.74	67.3%	-	-	3.2%	29.5%
Total metals	25.60	8.72	-	-	2.95	13.93	34.1%	-	-	11.5%	54.4%
Plastics	35.68	3.02	-	-	5.63	27.03	8.5%	-	-	15.8%	75.7%
Rubber and leather	9.16	1.67	-	-	2.50	4.99	18.2%	-	-	27.3%	54.5%
Textiles	17.03	2.51	-	-	3.22	11.30	14.7%	-	-	18.9%	66.4%
Wood	18.09	3.10	-	-	2.84	12.15	17.1%	-	-	15.7%	67.2%
Other materials	4.56	0.97	-	-	0.66	2.93	21.3%	-	-	14.4%	64.3%
Total materials in products	189.76	69.02	-	-	23.64	97.10	36.4%	-	-	12.5%	51.1%
Other wastes											
Food, other‡	63.13	-	2.59	17.71	7.55	35.28	-	4.1%	28.1%	11.9%	55.9%
Yard trimmings	35.40	-	22.30	-	2.57	10.53	-	63.0%	-	7.3%	29.7%
Miscellaneous inorganic wastes	4.07	-	-	-	0.80	3.27	-	-	-	19.7%	80.3%
Total other wastes	102.60	-	24.89	17.71	10.92	49.08	-	24.3%	17.3%	10.6%	47.8%
Total municipal solid waste	292.36	69.02	24.89	17.71	34.56	146.18	23.6%	8.5%	6.1%	11.8%	50.0%

* Includes waste from residential, commercial and institutional sources.
¥ Animal feed, bio-based materials/biochemical processing, codigestion/anaerobic digestion, donation, land application, sewer/wastewater treatment.
† Includes lead from lead-acid batteries.
‡ Includes collection of other MSW organics for composting.

Details might not add to totals due to rounding.
Negligible = Less than 5,000 tons or 0.05 percent.
A dash in the table means that data are not available.

Source: Advancing Sustainable Materials Management: 2018 Facts and Figures (2020).
Author: US Environmental Protection Agency.

Our MSW, or trash, is comprised of various items consumers throw away. These items include packaging, food, yard trimmings, furniture, electronics, tires and appliances. MSW does not include industrial, hazardous or C&D waste. Sources of MSW include residential waste, as well as waste from commercial and institutional locations, such as restaurants, grocery stores, other businesses, schools, hospitals and industrial facilities. Industrial facility waste includes waste from sources such as offices, cafeterias and packaging, but not process waste. Over the last few decades, the generation, recycling, composting, combustion with energy recovery and landfilling of MSW has changed substantially. Solid waste generation peaked at 4.74 pounds per person per day in 2000 and 2005, falling to 4.51 pounds per person per day in 2017. The higher rate of 4.91 pounds per person per day in 2018 reflects the change in food waste measurement methodology. The combined recycling and composting rate increased from less than 10 percent of generated MSW in 1980 to 35.0 percent in 2017. In 2018, the recycling and composting rate was 32.1 percent. Without including composting, recycling alone rose from 14.5 million tons (9.6 percent of MSW) in 1980 to 69 million tons (23.6 percent) in 2018. Although more tons were recycled in 2018 than ever before, the recycling rate decreased to the lowest levels since 2006. Composting was negligible in 1980, but it rose to 24.9 million tons in 2018 (8.5 percent). In 2018, for the first time in this report series, EPA revised its food measurement methodology to more fully capture flows of excess food and food waste throughout the food system. The resulting category, other food management pathways, accounted for 17.7 million tons (6.1 percent). Combustion with energy recovery was less than 2 percent of generation in 1980 at 2.8 million tons. In 2018, 34.6 million tons (11.8 percent of MSW generated) were combusted with energy recovery. Since 1990, the total amount of MSW going to landfills has increased by less than one million tons, from 145.3 million tons in 1990 to 146.2 million tons in 2018. The net per capita 2018 landfilling rate was 2.4 pounds per day, which was lower than the 3.2 per capita rate in 1990. (US Environmental Protection Agency 2018)

Figure 27—Evolution of MSW and Management (1960–2018)

Activity	1960	1970	1980	1990	2000	2005	2010	2015	2017	2018
Generation	88.1	121.1	151.6	208.3	243.5	253.7	251.1	262.1	268.7	292.4
Recycling	5.6	8.0	14.5	29.0	53.0	59.2	65.3	67.6	67.0	69.0
Composting*	neg.	neg.	neg.	4.2	16.5	20.6	20.2	23.4	27.0	24.9
Other Food Management**	-	-	-	-	-	-	-	-	-	17.7
Combustion with energy recovery†	0.0	0.5	2.8	29.8	33.7	31.7	29.3	33.5	34.2	34.6
Landfilling and other disposal‡	82.5	112.6	134.3	145.3	140.3	142.2	136.3	137.6	140.5	146.2

* Composting of yard trimmings, food and other MSW organic material. Does not include backyard composting.

** Other food management pathways include animal feed, bio-based materials/biochemical processing, codigestion/anaerobic digestion, donation, land application and sewer/wastewater treatment.

Details might not add to totals due to rounding.
neg. (negligible) = less than 5,000 tons or 0.05 percent.

† Includes combustion of MSW in mass burn or refuse-derived fuel form, and combustion with energy recovery of source separated materials in MSW (e.g., wood pallets, tire-derived fuel).

‡ Landfilling after recycling, composting, other food management and combustion with energy recovery. A dash in the table means that data are not available.

Source: Advancing Sustainable Materials Management: 2018 Facts and Figures (2020).
Author: US Environmental Protection Agency.

EPA analyzes MSW by breaking down the data in two ways: by material and by product. Materials are made into products, which are ultimately reprocessed through recycling or composting or managed by combustion with energy recovery facilities or landfills. They may also be processed by other management methods for food. Examples of materials that EPA tracks include paper and paperboard, plastics, metals, glass, rubber, leather, textiles, wood, food and yard trimmings. Products are what people buy and handle, and they are manufactured out of the types of materials listed above. Product categories include containers and packaging, nondurable goods, durable goods, food and yard trimmings. Containers and packaging, such as milk cartons and plastic wrap, are assumed to be in use for a year or less; nondurable goods like newspaper and clothing are assumed to be in use for less than three years; and durable goods, such as furniture, are assumed to be in use for three or more years. Some products, such as appliances, may be made of more than one material. Information about products shows how consumers are using and discarding materials and offers strategies on ways to maximize the source reduction, recycling and composting of materials. (US Environmental Protection Agency 2018)

Figure 28—Total MSW Generation by Material, 292.4 Million Tons in 2018

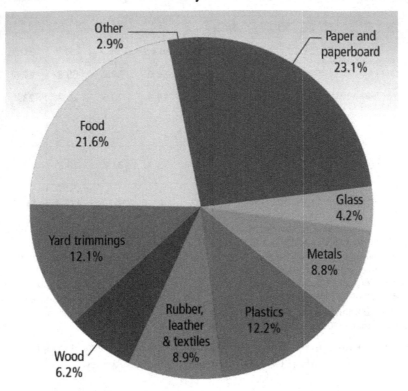

Source: Advancing Sustainable Materials Management: 2018 Facts and Figures (2020).
Author: US Environmental Protection Agency.

The year 2018 generated the most MSW ever recorded in the US history, at 292.4 million tons. Paper, food, and plastic made up the three largest contribution streams, with yard trimmings and rubber/leather/textiles tailing right behind.

Figure 29—Total MSW Landfilled by Material—146.2 Million Tons in 2018

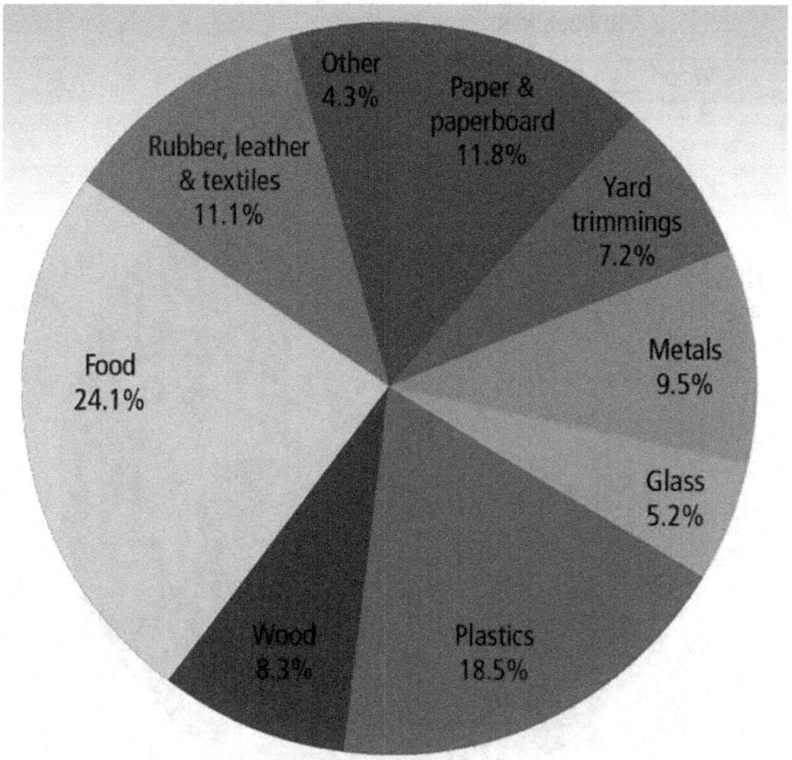

Source: Advancing Sustainable Materials Management: 2018 Facts and Figures (2020).
Author: US Environmental Protection Agency.

The amount landfilled in 2018 was 146.2 million tons, the highest amount recorded in the last ten years. Population in that year also reached an all-time high: 325 million. About sixty-nine million tons of materials were recycled; however, only 4.4 percent (about three million tons) was plastic material. Paper and paperboard contributed over 66 percent (about forty-six million tons) of the material that was recycled. Metals made up 12.6 percent (about nine million tons) of materials recycled.

Figure 30—Total MSW Recycled by Material, Sixty-Nine Million Tons in 2018

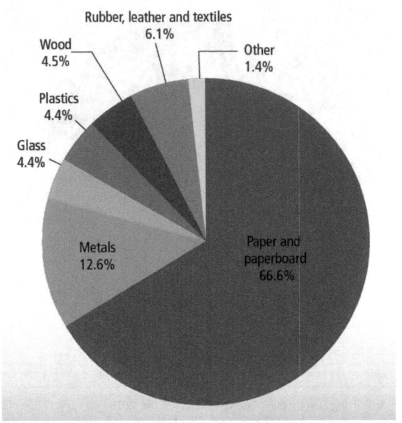

Source: Advancing Sustainable Materials Management: 2018 Facts and Figures (2020).
Author: US Environmental Protection Agency.

The big eye-openers from that US EPA data are the increases of plastic and food waste, as well as total waste increasing year after year in landfills since 2005. In 2018, over twenty-seven million tons of plastic, and over thirty-five million tons of food waste went to landfills. Only three million tons of plastic was recycled, and 2.59 million tons of food waste was composted. These facts could lead one to believe there is tremendous opportunity for these two material types alone for smarter and more efficient recycling and composting solutions. What is somewhat startling is that total waste generation increases year after year and reached an all-time high of over 292.4 million tons in 2018, as the population reached an all-time high of 325 million. Does that mean that as the population grows, as it is predicted to do, so will waste? It is a valid question to consider.

The amount of waste that is being recycled and composted annually has, in fact, increased, but the issue is that it is increasing very slowly compared to population growth and economic growth. The opportunity to recycle more materials, like plastic, is enormous. However, it will take an ambitious and dedicated effort by stewards to come up with new, efficient solutions and systems to better use and reinvent waste. There is no doubt that great opportunity lies ahead. Environmentally, we are making big strides already: "In 2018, about 94 million tons of MSW in the United States were recycled and composted, saving over 193 MMTCO2E (GHG benefits). This is comparable to the emissions that could be reduced from taking almost 42 million cars off the road in a year" (US Environmental Protection Agency 2018). If we doubled the amount of recycling and composting, the environmental impacts would be immense. If you consider the economic benefits from recycling and composting efficiency too, it would be an environmental plus an economic win for civilization. If you are brave enough to tackle the construction and demolition debris dilemma as well, and the 144 million tons landfilled in 2018, you are well on your way to helping change the world.

Michigan's Waste Stream and Sustainable Management Efforts

Michigan's waste stream looks a little different compared with the average state. Around 20 percent of its waste comes from Canada. Michigan made up about 6.5 percent of the entire MSW generated in the United States, according to the EPA 2017 report.

2017 MSW Generation Stats:

- Michigan waste: 12,574,640 tons
- Canadian waste: 3,524,307 tons
- Other states: 769,282 tons
- Total waste disposed in Michigan landfills: 16,868,230 tons

Data Source: Michigan Department of Environmental Quality (DEQ), 2017 Landfill Report.

**Figure 31—Solid Waste Disposed in Michigan Landfills
in 2017 (All Figures in Cubic Yards)**

Fiscal Year	Michigan	Canada	Other States	Totals
FY 2008	39,913,636	10,722,164	6,484,096	57,119,896
FY 2009	34,751,365	9,054,371	4,031,983	47,837,719
FY 2010	34,802,866	8,757,014	2,563,056	46,122,936
FY 2011	35,857,919	6,983,127	2,886,955	45,728,001
FY 2012	34,485,534	6,512,223	2,912,750	43,910,507
FY 2013	34,233,158	7,594,716	2,785,700	44,613,574
FY 2014	36,281,718	7,639,167	2,830,235	46,751,120
FY 2015	36,862,512	8,090,942	2,672,899	47,626,353
FY 2016	37,488,887	8,883,958	2,690,793	49,063,638
FY 2017	37,723,925 [4]	10,572,922 [5]	2,307,845 [6]	50,604,692 [7]

Source: Michigan Department of Environmental Quality (DEQ), 2018 Landfill Report.

Recycling represents a significant opportunity in Michigan, as only 15 percent of the waste was recycled in 2014, which is below the national average. Michigan has a tremendous opportunity to be more aggressive in terms of advancing in the realm of recycling. In 2014, 1.54 million tons was recycled, and of that, only 5 percent was plastic. Paper and organics made up 31 percent and 29 percent respectively, which made up most of the recycling streams. Based on these numbers, there is great potential for individuals and organizations to implement new solutions and systems for repurposing and reinventing waste. Michigan is currently behind the curve in terms of national recycling efficiency and percentage. The good news is that there are motivated stewards that are currently helping Michigan to evolve and innovate in the field of sustainable waste management.

Figure 32—Michigan Recycling Rate, 15.3 percent (2014)

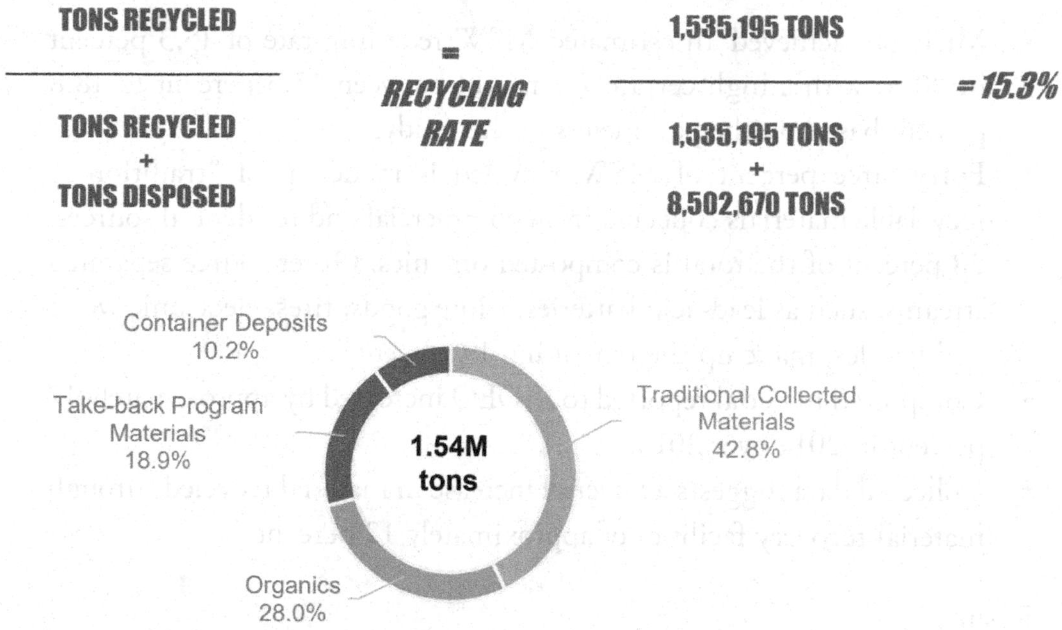

$$\frac{\text{TONS RECYCLED}}{\text{TONS RECYCLED} + \text{TONS DISPOSED}} = \textbf{RECYCLING RATE} = \frac{1,535,195 \text{ TONS}}{1,535,195 \text{ TONS} + 8,502,670 \text{ TONS}} = 15.3\%$$

Container Deposits
10.2%

Take-back Program
Materials
18.9%

1.54M tons

Traditional Collected
Materials
42.8%

Organics
28.0%

Source: Michigan Department of Environmental Quality, Measuring
Recycling in the State of Michigan, 2014 Recycling Rate.

Overview:

In 2014, the State of Michigan achieved a total statewide MSW recycling rate of 15.3 percent. Of the total amount of material recycled, only 43 percent is composed of 'traditional' recyclable materials collected from commercial and residential sources. Twenty-eight percent of the total is composted organics, mostly yard waste. The container deposit program accounts for 10 percent, and other source separated streams (such as lead-acid batteries, white goods, tires, e-waste, and textiles) make up the remaining 19 percent. The MRI survey process followed the recommended survey guidelines established by the US EPA for measuring recycling rates, and submitted data was applied directly to the respondent communities. Additionally, data received was used as the basis for an extrapolation of recycling activity to gap communities which have analogous and relevant demographic characteristics that are likely to be reflected through recycling performance. The types of data collected and the data collection methods are described below.

Highlights:

- Michigan achieved an estimated MSW recycling rate of 15.3 percent in 2014, with a high certainty range of between 13.3 percent to 18.8 percent based on the parameters in this study.
- Forty-three percent of MSW recycled is made up of "traditional" recyclable materials collected from commercial and residential sources. 28 percent of the total is composted organics. Other source separated streams, such as lead-acid batteries, white goods, tires, electronic waste and textiles, make up the remaining 19 percent.
- Composted material reported to MDEQ increased by approximately 17 percent in 2014 over 2013.
- Collected data suggests an overall increase in material recycled through material recovery facilities of approximately 12 percent.

Results:

- Information sharing was voluntary and the project team received data from nine private and public entities operating 13 Michigan mrfs. This participation represents Michigan's largest recycling facilities and includes approximately 40 to 60 percent of the total throughput from material recovery facilities in the state of Michigan.
- Container deposits accounted for 10.2 percent of 2014 recycled MSW, or 1.6 percent of total MSW.
- In 2013, the volume landfilled was 21,581,275 cubic yards, and this volume increased by 5.3 percent in 2014 to 22,715,636 cubic yards. In 2013, 881,953 tons of material was incinerated, Traditional Collected Materials 42.8 percent Organics 28.0 percent Take-back Program Materials 18.9 percent Container Deposits 10.2 percent. 1. TONS RECYCLED + TONS DISPOSED = 1,535,195 TONS + 8,502,670 TONS = 15.3 percent = RECYCLING RATE, 8 increasing by 11.9 percent to 986,660 tons in 2014. If disposed volumes were to have remained steady year over year, the overall recycling rate would be 16.1 percent.
- In 2014, 3,459,241,584 containers were redeemed as part of the Michigan container deposit scheme, yielding 161,387 tons of high-value materials including plastics, metals and glass. This is relatively consistent relative to 2013.
- Composted material was approximately 441,843 tons in 2014 (Michigan Department of Environmental Quality 2016).

Figure 33—Recycling Stream by Material (2014)

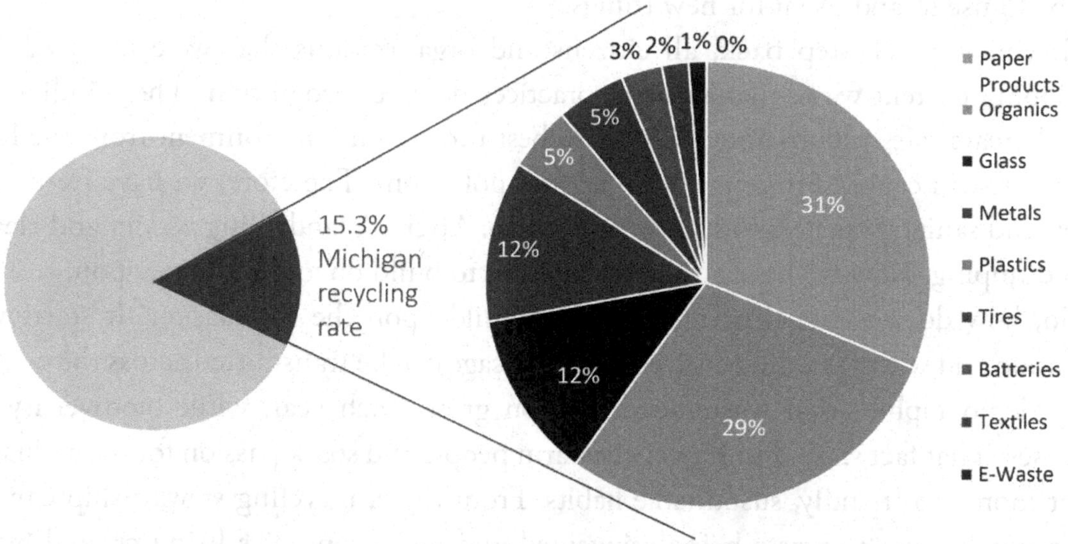

15.3% Michigan recycling rate

3% 2% 1% 0%
5%
5%
12%
31%
12%
29%

- Paper Products
- Organics
- Glass
- Metals
- Plastics
- Tires
- Batteries
- Textiles
- E-Waste

Source: Michigan Department of Environmental Quality, Measuring Recycling in the State of Michigan, 2014 Recycling Rate.

Potential Future of Sustainable Waste Management

There is a major dichotomy in the United States as waste generation (particularly MSW) is increasing, while biodiversity loss is growing. The big picture is that we humans are using too much of the Earth's resources too fast, landfilling more waste each year, and polluting the Earth more each day. We are nudging Earth toward a very dangerous path where consequences are unknown. Biodiversity loss, global pollution, increased carbon footprint, climate change, and wasting valuable materials and resources are human impacts on the planet that are wreaking havoc, and we currently lack the innovation to best use waste to generate new economic opportunities.

Human activity creates many forms of waste, and we may never be able to completely rid ourselves from it. However, what we can do is implement new strategies to divert much less waste to landfills and come up with more intelligent solutions to recover it. Waste is not yet seen as a valuable resource, at least on a large scale. If it were considered highly valuable, there would be much less waste going to landfills every year. The main problem that needs swift reversal is the overconsumption of resources and growing biodiversity loss. We need almost 1.7 more Earths to keep up with the current human demand, and that is not sustainable. We currently have no other planets to use for resources, so our best chance for sustaining life on Earth is to implement smarter resource consumption and learn how to better use and reinvent waste. There is hope, and humans can team together to create the positive change that is necessary to save our planet. The positive change will not only literally save the planet but will also create new economic opportunities around sustainable waste management. The more waste we recover as a nation, the more opportunity there will be to use it, and use it for new things.

Taking a quick step back, all citizens and organizations that were involved with developing current waste-management practices deserve recognition. These individuals helped create safe systems and practices to best protect our environment from the harm that waste can cause (dumping, littering, and pollution). Therefore, we have fresh food, water, and sanitary shelters to survive and thrive. After all, landfilling is safer and cleaner than dumping. Now we have a solid foundation to build on and improve upon.

So, how do we start this transition and build upon the foundation? It starts with education and with spreading awareness. A message can be transmitted across the country to all demographics that waste accumulation grows each year, while biodiversity loss increases. That fact should immediately alarm people and spark passion for individuals to adopt more eco-friendly, sustainable habits. From there, recycling stewardship can and will grow, leading to waste being reinvented and new economies being created by the philosophy of sustainable waste management. History reminds us that we have already made significant strides and implemented key strategies and infrastructures to safely manage waste, protect our environment, and to start recycling and reusing waste. We have evolved from uncontrolled dumpsites and littering/pollution to controlled landfilling, recycling and reusing, composting, and anaerobic digestion, to WTE technology. All these efforts have laid a strong foundation to build on, especially as technology has advanced and human ingenuity improves rapidly. The future is filled with opportunity and great potential; we just have to find the light of positivity and learn to stay on that path.

Tips for Stewards: Fourteen Solutions for Best Recovering Waste

1. Find methods and/or organizations to collect or recover all waste: reuse or repurpose, recycle, compost, use anaerobic digesters, incineration, or waste-to-energy and landfill.

2. Respect Earth's resources and use all materials sparingly. When you accumulate waste or materials that you're not sure how to dispose of, contact your local city or county government to learn the best methods for waste disposal.

3. Never litter or pollute. If you do, make yourself aware of the situation and change practices.

4. Educate yourself on current waste-recovering methods (above section) in the United States, both locally and nationally. Try to implement these methods into your practices.

5. Tour a landfill, recycling center, compost center, or waste-to-energy plant to see how it looks aesthetically, and learn how they operate. It can be motivating.

6. Implement your own detailed best-practice plan to reduce and best recover waste at your home, business, or organization. Measure your progress each year and create new goals.

7. Develop recycling stewardship programs at your home, business, or organization, and promote sustainable waste management measures.

8. Help start a national or global movement to name waste as the least valued resource available to humankind. Educate people on the number of landfills in the country.

9. Spread the word about how biodiversity loss is increasing, while waste (MSW) accumulation grows year after year, reaching an all-time high of 292.4 million tons in 2018.

10. Join organizations and teams that practice and promote sustainable waste recovery habits. Volunteer, support the mission, and help these organizations and teams grow stronger.

11. Continually learn, innovate, and implement new methods for reducing waste and ways to reinvent waste. There are always new ideas that can make a difference.

12. As the world changes and evolves, learn to be flexible and moldable, just like Earth.

13. Continually sharpen your skills to be a tactical leader. Listen to others and learn from them. As people work together, positive change can grow more quickly and more efficiently.

14. We should continue to unite on the issues that arise. Only together can we make a lasting change that will save the Earth.

Step 3—Reinvent

Reinvent Waste as the Most Undervalued Resource/Material on Earth

Undervaluation of Waste

Most people would agree that it's cheap and convenient to send MSW to landfills. And is it convenient and easy for individuals to dispose of waste through methods such as littering and pollution too. As discussed earlier in step 2, recover, waste streams are continuing to grow in millions of tons every year in landfills, dumpsites, and generally across the globe through littering and pollution. But why? Is there a lack of education and awareness about waste and its negative impacts on our environment? Or maybe it's that the economic benefit of waste production and management is so strong that companies, organizations, and governments are afraid of making proactive changes. What could be the underlying issues or fundamental flaws that encourage waste production rather than reduce it? Let's explore how producing and disposing of waste is inexpensive and potentially not valued correctly.

First, the economic impacts of waste and recycling are significant. Simply put, the modern waste and recycling industries have been built up and are now an important part of our economy. Early on, waste management was just a response to deal with negative environmental impacts through dumping and pollution. As awareness and education grew, collective minds together transformed waste management and created innovative ways to reduce, reuse, and recycle. According to the National Waste and Recycling Association (NWRA), in 2016, about 384,000 people worked in the waste and recycling industry, and total direct revenue hit almost ninety-four billion dollars in the United States (National Waste & Recycling Industry 2016). If you include indirect and induced measures, those numbers would be much larger. The numbers also vary greatly from state to state based on the population and other factors like consumer spending and the number of businesses producing in that state. For instance, California claims that almost forty-five thousand individuals work in the waste and recycling industry, producing about thirteen billion dollars in economic impact. If indirect and induced measures were considered, employment would equal 143,000 individuals, producing $26.5 billion in revenue (National Waste & Recycling Industry 2016). If you compare this to Michigan, direct employment and total revenue were eleven thousand people and almost $2.9 billion, respectively. Including indirect and induced measures equates to almost thirty-six thousand individuals employed in this industry and about $5.9 billion in revenue. So, depending on how one looks at the effects of waste generation, it could be seen as a benefit. However, as we have discussed, there are severe negative impacts occurring now to the environment, and that can be solved. As waste accumulation continues to grow, it would make sense to use the waste rather than to bury it. Shouldn't we use the true value of the waste?

There are many things trending upward in parallel, however, and not all are creating positive impacts. Although constructive waste and recycling management is growing, so too is waste accumulation and environmental damage (more waste is going to landfills, and various contaminations and pollution like microplastics are harming the ocean ecosystems). Also, the ecological footprint, population, consumer spending, and overall GDP are growing in the United States and across the globe. Those factors are continuing to play a role in the evolution of three major impacts, shown in Figure 34. Those impacts are environmental, waste accumulation, and economy of waste management. The growing economic impact of waste management is positive; however, the environmental damage and increase of waste accumulation are indicators that change is necessary. That need is an increase in sustainability-focused efforts toward waste management and innovation. Hopefully these efforts of reinventing waste will create a much-needed balance so that civilization can live in harmony with the natural world. Reinventing waste should spur further economic growth and new sustainable economies. Reinventing is rethinking, reimagining, renovating, and revaluing. Reinventing is innovating and inventing a new idea, a new system, and a new way to think about sustainable waste management. Reinventing waste will help us properly value it moving into the future.

Figure 34—Evolution of Impacts, Positive (+) versus Negative (-)

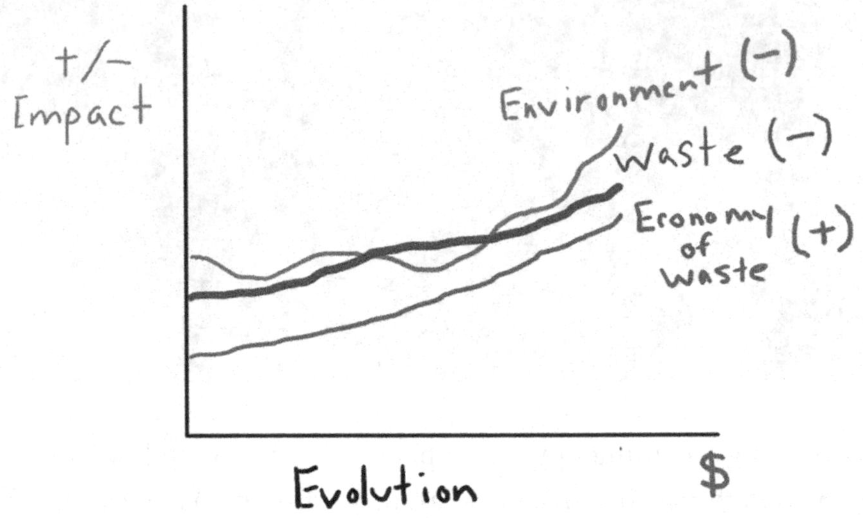

+/–
Impact

Environment (–)

Waste (–)

Economy
of
waste (+)

Evolution　　　$

Source: Subjective visualization by Tyler Kanczuzewski.

Is waste undervalued? Tipping fees is an interesting topic that should be considered for reinvention measures. Landfills charge what is a called a tipping fee. A tipping fee is a charge per metric ton of waste dumped at a landfill. A simpler way to understand a landfill is that it is a business, and its product is selling space. The space is then used for trash. The more waste a landfill can take, the more it can charge, likely leading to more revenue and profits. Most landfills constitute prime real estate and have overhead costs (costs to run a facility). So, for them to remain open and turn profits, governments, organizations, people, and businesses are charged tipping fees for disposing of their waste at the landfill. Landfills and waste-retrieval services, or waste-management companies, are mostly owned by public and privately held organizations, as well as by government institutions. Some regional infrastructures are a combination of all the above, but they are all operating to cover overhead and trying to make profits for investors. Tipping fees are the product that a landfill sells to make money. The EPA has tracked average tipping fees across the country since 1980 (see Figure 35). The average tipping fee in 1980 was $19.82, and in 2015, it was $48.10. However, in the mid-1990s, the fee hit an average of $50.06. Basically, for the last twenty years, the fee has remained steady. However, there has been an uptick in fees nationwide over the last several years. Why is this? One possibility is that landfills are profitable and the return on investment (ROI) is strong. Another possibility is that supply (landfill space) and demand (volume of waste) are both strong and growing constantly, leading to an economic equilibrium. However, factors likely not being considered are a growing pollution, lack of ecologic resources, and increasing environmental damage. Individuals and stakeholders in the waste industry should start treating waste differently and realize alternative values. It can't be just about revenue from tipping fees and making landfills bigger so that they can consume more waste and potentially make more profits.

Figure 35—Average US Tipping Fee for Landfills

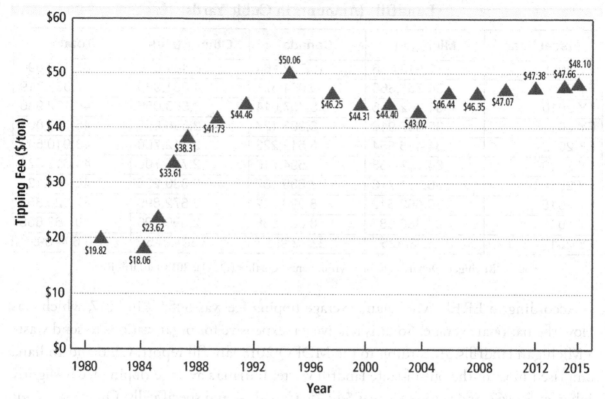

Source: US Environmental Protection Agency, Advancing Sustainable Materials Management Fact Sheet, 2015.

The Environmental Research and Education Foundation (EREF) keeps track of 1,540 active landfills, both public and private, to analyze tipping-fee rates and changes in the United States. Tipping rates vary based on region, as well as landfill operation size. Connecticut charges two hundred dollars per ton, which is the highest rate, and Mississippi charges a low of $21.67 per tonnage as of 2017 (Environmental Research & Education Foundation 2017). Why do fees vary greatly? Space might be extremely limited in parts of the country, and landfills are being maxed out, while available land for new landfill centers is becoming limited. Land value in certain parts of the country is prime, high in demand, and mainly based on variables like population, geography logistics, and industry development. Could it also be an incentive for disposal companies to send waste out of state or potentially to initiate more recycling programs within the state? The situation in Michigan brings up many interesting questions and topics for discussion. According to the Michigan Department of Environmental Quality (MDEQ), the state landfills (roughly forty-four) consumed almost seventeen million tons of waste in 2017. Interestingly, 21 percent (3.5 million tons) of that waste came from Canada, and almost 5 percent (769,000 tons) came from other US states (Michigan Department of Environmental Quality 2018).

**Figure 36—Solid Waste Disposed in Michigan
Landfills (Amounts in Cubic Yards)**

Fiscal Year	Michigan	Canada	Other States	Totals
FY 2008	39,913,636	10,722,164	6,484,096	57,119,896
FY 2009	34,751,365	9,054,371	4,031,983	47,837,719
FY 2010	34,802,866	8,757,014	2,563,056	46,122,936
FY 2011	35,857,919	6,983,127	2,886,955	45,728,001
FY 2012	34,485,534	6,512,223	2,912,750	43,910,507
FY 2013	34,233,158	7,594,716	2,785,700	44,613,574
FY 2014	36,281,718	7,639,167	2,830,235	46,751,120
FY 2015	36,862,512	8,090,942	2,672,899	47,626,353
FY 2016	37,488,887	8,883,958	2,690,793	49,063,638
FY 2017	37,723,925 [4]	10,572,922 [5]	2,307,845 [6]	50,604,692 [7]

Source: Michigan Department of Environmental Quality (DEQ), 2018 Landfill Report.

According to EREF, Michigan's average tipping fee was $45.97 in 2017, which was below the national average. So, it is relatively inexpensive for organizations to send waste to Michigan landfills. According to the MDEQ 2018 landfill report, Ohio and Indiana comprised most of the out-of-state landfill waste. Indiana's average tipping fee is slightly higher, at $49.67, and Ohio's fee was $44.75. Canada—and specifically, Ontario—sends a good portion of its waste to Michigan, according to the MDEQ. Ontario is practically a neighbor of Michigan just as other Midwestern states are. It is interesting to realize that another country is shipping waste across borders to dispose of it. It is likely that Canada sends waste across borders because it's cheaper to dispose of in Michigan. And it is likely that Ohio and Indiana also send trash to Michigan because, primarily, it is cheaper, and secondarily because of logistics or space constraints.

Could tipping rates be adjusted nationally to create more sustainable waste management systems? Could that be a simple fix to greatly reduce the amounts of waste going to landfills, all while contributing to a growing economy focused on waste and recycling management? According to Dar Baas, director of Kent County Department of Public Works, landfill costs (or cost delta) versus recycling costs is too expensive. Baas believes it's cheaper for users to get rid of waste in a landfill than to spend money on recycling efforts (Bass 2018). Nationwide recycling efforts are increasingly growing, as seen from data in step 2, recover. However, on average, it is cheaper to send waste to a landfill than to pay for or invest in recycling systems to recycle or repurpose waste. Is that a problem, and should it be cheaper to recycle than to send waste to landfills? It would be an interesting change and a big economic shift if landfills were to increase tipping rates to encourage recycling efforts. Couldn't landfills reinvent themselves and change their business models to be more involved in recycling or partner with recycling organizations? If recycling could become cheaper than landfilling, it would seem highly likely that the waste and recycling industry would continue to grow and spur new innovations. New recycling innovations would play a part in the reinventing of waste and lead to more opportunities. Imagine what you could do with vast piles of resources that would otherwise be buried in the ground. Are we willing to invest effort in reinventing waste?

A New Mindset Shift—Stewards of the Four-R Earth

Waste is currently not being used as a resource, and it is undervalued. The growing accumulation of waste is a key indicator that we have not been fully sustainable as a nation and have not taken the opportunity to fully tap into the value of waste. MSW—and particularly product and packaging waste (especially in plastic)—is growing fast. Waste volume in general is growing at unprecedented rates, which calls for newer and bigger landfills. Not only does the United States, and states like Michigan and cities like New York, not have space for these enlarged or new landfills, society can't afford the environmental harm, economic loss, and deterioration of human health. In Kent County, Michigan, landfill space is becoming extremely limited. Some of the waste is being sent to a WTE plant, and more waste is being sent to recycling centers to implement more systems to reuse waste. However, the county is now dealing with uncertain times as space is becoming severely limited for a new landfill.

According to the EPA in 2018, the United States generated 292.4 million tons of MSW, which is a new record high. More than 50 percent (146.2 million tons) of that waste went to landfills, and only 23.6 percent (69 million tons) was recycled; the other 26.4 percent (77.2 million tons) was composted, combusted at WTE facilities, or alternatively managed. That means that over 50 percent of the MSW waste in 2018 was not efficiently used. And that doesn't include the 144 million tons of construction and demolition debris landfilled. This has been the common theme year after year, even though there have been some improvements along the way.

As discussed earlier, it is generally cheaper to send waste to landfills versus trying to recycle. Studies are now showing that the United States is living out of balance with the natural world and consuming resources faster than they can regenerate, which is not sustainable. According to our ecological footprint (rate of consuming resources), the United States needs five Earths, whereas the entire world needs roughly another 1.7 Earths to keep up with modern demand. All the while, waste and pollution are increasingly growing. Because of this dichotomy, it would make sense to value waste much differently. Waste can be used and reinvented as a valuable resource to help save the planet as well as to create new business platforms and economies focused on stewardship and sustainability, and a new waste management system.

The United States has indeed progressed toward more sustainable waste management, as step 2, recover, discussed. There have been numerous movements, initiatives, and policies that have been implemented to spread awareness across the country to get people to think differently about resource conservation and recovery. One of those initiatives was the three Rs mindset adopted in the 1970s: *reduce, reuse, recycle*. The three Rs mindset worked well and raised awareness about waste. It was so powerful that it is still used today. The 1970s was a pivotal decade for creating a sustainable mindset shift to fight the growing epidemic of waste and pollution. Now, about forty-plus years later, sustainability efforts have grown, and waste management is improving slightly each year. However, are these efforts as progressive as they could be, and can sustainable waste management be reinvented? The answer is yes, and we can greatly reduce the amount of waste that is going to landfills, litter that is suffocating our lands, and waste that is polluting our precious waters by adopting a new and shifting mindset.

The new mindset is a four-step plan: respect, recover, reinvent, and restore (four-*R* Earth). The four-step plan will help establish a new mindset to respect the natural world, recover waste to the best of our abilities, reinvent waste to use it as a valuable resource, and restore the Earth so that we live in a more balanced way and assure that Earth provides for all future generations. Awareness has already started to spread throughout the country about waste accumulation growing and pollution increasing. Waste from single-use disposables, packaging, and plastic is arguably growing too rapidly. New initiatives, movements, and policies have been recently implemented across the country and throughout various organizations. A foundation for improved sustainability is being laid as we speak, and the four-step plan will add to that foundation and give birth to a mindset shift that will reinvent waste and progress the sustainable waste management movement.

The capstone of the new mindset and four-step plan is reinventing waste. Step 3, reinvent, addresses the severe problem that the United States, states like Michigan and heavily populated regions, and most of the world are living out of balance with the natural world. This is all while we are burying and not properly valuing waste. Now is the time to explore ideas and methods that any person, steward, or organization can implement to reinvent waste, to help save the Earth from potential environmental harm, and to realize a higher value in waste streams. Last, step 4, restore, closes with the idea that if proper steps are taken, humans can regain a sense of balance with Earth and restore it to become more sustainable for generations to come.

We only have one Earth, so we need to try our best to live in balance with it. The Earth is resilient, and if we respect it, we can restore it so that it provides life for future generations. By reinventing waste, we are being good stewards for Earth. This study, and especially step 3, reinvent, raises awareness and provides ideas and mindsets to live by for individuals who want to be good stewards. This study in general is a steward's guide for reinventing waste. If we don't reinvent waste, we will continue to negatively impact the natural world, which could be an extremely dangerous road that humanity should not venture down. By reinventing waste, we can transform the waste industry and create more superior sustainable economies than ever before.

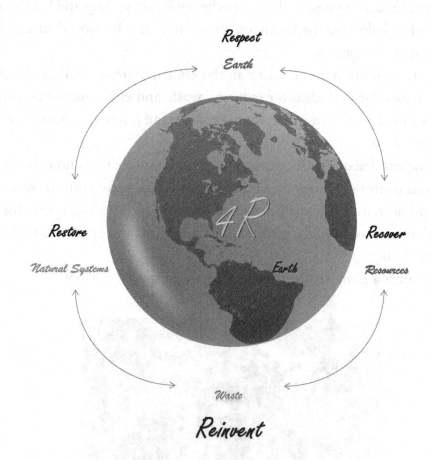

Respect
Earth

Restore
Natural Systems

4R

Recover
Resources

Earth

Waste

Reinvent

Goal Ideas for Step 3, Reinvent Waste:

- Identify the benefactors from the reinventing waste revolution.
- List movements, companies, and organizations that are already reinventing waste.
- Promote case studies that have been successful.
- Create and raise awareness and educate the public about why and how waste can be reinvented to help save the Earth and create new and improved sustainable waste management economies.
- Help reduce pollution and littering in the United States as well as globally.
- Spread knowledge and ideas for reducing waste and environmental harm.
- Provide strategies for stewards to manage waste and improve processes to minimize and repurpose waste.
- Determine eco-friendly last-resort options for waste that has no other use/home.
- Create and form new innovative economies that properly reinvent waste.
- Invent, pioneer, and provide innovative ideas for reinventing waste; for instance, sustainable processes, infrastructures, business ideas, and platforms to reduce waste or repurpose it.
- Help establish a more realistic value for waste, especially MSW.

Reinvent your Waste, but not like this

Value Waste Correctly to Solve a Potential Crisis

Reinventing waste can benefit anyone, especially those who want to be good stewards. If we treat the planet like it's all we have, more sustainable decisions will be made every day, but it's challenging to spread awareness and get everyone thinking that way. As the Golden Rule says, "Treat others as you want to be treated." It's about time we treat the Earth how we would want to be treated. One way to treat the Earth better is by reinventing waste. Doing so will create economic benefits. By reinventing waste, we can potentially solve and reverse three major issues that are occurring in the United States as a whole, in states like Michigan, and in various populated regions and the planet.

1. Waste accumulation, littering and pollution are increasing and leading to major ecological issues. (*Stop wasting waste*). These patterns will likely continue at increasing rates unless something is done. Factors that contribute to these trends are population growth, resource extraction, bypassing logistical hurdles, and consumers consuming more products, goods, and services. Increases in waste is wreaking havoc on the natural world. Refer to step 1, respect.

2. Earth's resources are being extracted or consumed too rapidly. (*Stop consuming and developing so rapidly; instead, develop and consume smarter*). Earth cannot regenerate at the rate we are consuming. We could slow down or get smarter with our consumption of resources so that Earth can provide for future generations. We must respect the available resources and consider using resources that are more prevalent, more renewable, and more sustainable. Refer to step 1, respect, for more on this issue.

3. Waste is not being valued properly, or not valued at all. *(This provides an economic opportunity for us all)*. Waste is piling up in landfills, and littering and pollution are not slowing down. Newer forms of waste management exist. The problem is that these methods are not valuing waste properly yet, or not at all. The only value seen now in waste is in the infrastructure involved in transporting and managing waste and the economic impact that has. There is an abundance of waste. Now it's time for people to separate and collect streams in which they find value.

Do you ever think, "Should the economy and human development slow down or stabilize?" In this age, the modern Digital Revolution, things seem to be moving fast, and in many cases, it's true. The Digital Revolution, as with the Industrial and Technological Revolutions, have accelerated human development (resource or biodiversity extraction) faster than ever, historically speaking. Logistical hurdles have been overcome to improve resource extraction, mass production, cost and convenience, and delivery. This movement of overcoming logistical barriers has spurred growth in capitalism, business competition, and consumerism. This movement has given birth to our modern-day globalized economic engine, an engine that provides seemingly unlimited resources that are available at the click of a button. The key problem, as discussed, is that we know we do not live in a world of unlimited resources. We know that waste is piling up quickly, that it's not being fully used or valued, and that it's harming the environment. The good thing is there is always hope. Right now, there are good stewards working hard every day to combat waste and are finding ways to value it more efficiently and properly. We will now discuss some of these movements and the collective efforts toward make a more sustainable world.

Before digging into all the ideas that have already been implemented to reinvent waste, it would be beneficial to put a value or dollar amount on waste, particularly MSW. The type, value, and amount of waste can vary from region to region, and from state to state. For instance, Michigan threw out over sixteen tons of waste, compared with 267.8 tons nationally in 2017 (that does not include litter and pollution). A study was recently done by the West Michigan Sustainability Business Forum (WMSBF) that put a value to certain waste streams.

2016 WMSBF Report on Economic Impact Potential and Characterization of MSW:

> This project was funded primarily through a $50,300 grant from the MDEQ. In his 2012 special message on energy and the environment Governor Rick Snyder acknowledged the low recycling rate in Michigan and committed to creating a plan to improve that rate. In response, the Michigan Department of Environmental Quality (MDEQ) convened a stakeholder workgroup to begin a dialog to advance recycling in Michigan. One finding of that group was a need for more data and information to inform state and local decision makers.
>
> This study provides information and analysis on the composition of municipal solid waste currently landfilled and incinerated in Michigan, and the economic value of this material. Its findings are derived entirely from field studies, verifiable market prices for recycled commodities, and peer-reviewed academic studies.
>
> A coalition led by West Michigan Sustainable Business Forum sampled nearly 10 tons of garbage from eight sites throughout Michigan. Working with technical consultant Fishbeck, Thompson, Carr, & Huber, Inc., WMSBF has created a waste characterization report for local communities and the state, providing much-needed data to decision makers on the materials sent to Michigan landfills and incinerators. That information and commodity pricing data provided by WMSBF member companies allowed Grand Valley State University to perform an analysis of the potential economic impact to the state.
>
> WMSBF partners opening their facilities to the project include Republic Services, Kent County Department of Public Works and Muskegon County. Members that provided commodity information and other support include Rapid Green Group, PADNOS, Valley City Electronic Recycling, Organicycle, New Soil and My Green Michigan.

- Study sampled waste from eight sites across Michigan, sorting approximately 10 tons of material
- Michigan garbage contains an estimated $368 million of recyclable material
- Capturing this material would have been $399 million in economic impact, or an estimated 2,619 jobs
- West Michigan garbage contains an estimated $52 million of recyclable material
- In communities with recycling programs, 42 percent of garbage is "easily recyclable"
- Food waste accounted for 13.6 percent of garbage, the largest source of divertible material
- Corrugated cardboard [is] a "high-volume, high-value opportunity" material at 8.4 percent, but more prevalent in commercial waste (10.5 percent commercial to 5.8 percent residential)

The study concludes that efforts to increase the recycling rate in Michigan should first focus on the 42 percent of materials that have market value, which would include all standard recyclable commodities but glass, plus textiles. To achieve the stated goal of doubling the Michigan recycling rate to 30 percent, the state must increase the quantity of diverted material by approximately 1.5 million tons per year through a combination of recovery and source reduction, according to the study. To accomplish this, it offered the following recommendations:

1. Aggressively promote efforts to increase recovery of corrugated cardboard, prioritizing commercial audiences.
2. Support efforts to increase availability and usage of conventional recycling programs with a goal to increase recovery of noncorrugated paper products, metal, and high-value plastic resins HDPE and PET.
3. Through recovery or source reduction, decrease the quantity of electronic waste disposed of in Michigan landfills by half.
4. Promote source reduction and diversion of food waste.
5. Promote source reduction of low-value plastic resins.
6. Initiate efforts to increase recycling channels for textiles and promote availability of textile recycling.
7. Educate the public on the financial difficulties of recycling and waste diversion.
8. Pursue opportunities for further study.

The study also highlighted unique findings regarding electronic waste, deposit bottle containers, yard waste and textile recycling. In addition, the study includes regional reports for West Michigan, Kent County and Muskegon County, which dispose of an estimated $52 million, $27.8 million and $7.2 million worth of recyclable material each year, respectively (West Michigan Sustainability Business Forum 2016).

Figure 37—Michigan Municipal Solid Waste Composition (Mean % by Weight), 2016

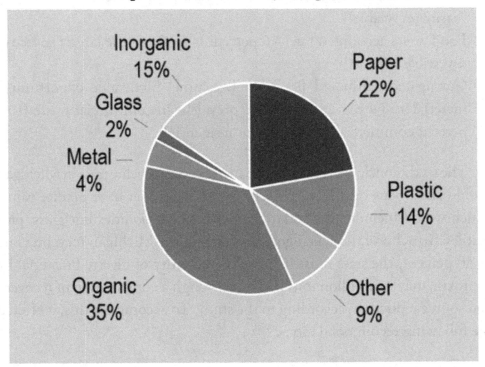

Source: West Michigan Sustainable Business Forum, 2016 Michigan MSW Valuation Study.

Figure 38—Total Value of Michigan MSW Commodities Disposed ($), 2016

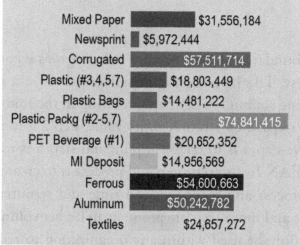

Commodity	Value
Mixed Paper	$31,556,184
Newsprint	$5,972,444
Corrugated	$57,511,714
Plastic (#3,4,5,7)	$18,803,449
Plastic Bags	$14,481,222
Plastic Packg (#2-5,7)	$74,841,415
PET Beverage (#1)	$20,652,352
MI Deposit	$14,956,569
Ferrous	$54,600,663
Aluminum	$50,242,782
Textiles	$24,657,272

Source: West Michigan Sustainable Business Forum, 2016 Michigan MSW Valuation Study.

Across the country in 2018, the United States recycled and estimated 23.6 percent of total MSW, according to the EPA. In 2014, Michigan reported only a 15.3 percent recycling rate (Michigan Department of Environmental Quality - Recycle, 2016). Imagine if Michigan were to use the value of the waste from 2016. Figure 37 estimates a total value of about $368 million in revenue if that waste were collected and repurposed in some shape or form. In this region alone, there was over seventy-four million dollars in plastic (#2–5, 7) packaging value. This is just the total value of the waste streams; it does not include pollution impacts or other waste streams. Imagine the new economies related to sustainable waste management that could be created by capturing this value! By reinventing waste, new value will be created and will spur new green economies and industries that are far more impactful both in terms of revenue and leading to a more sustainable planet.

Modern Movements and Initiatives That Do and Can Value Waste

Recent years have seen great waves of innovation, movements, and new standards adopted. These waves are helping raise awareness and are offering potential solutions for good stewards to reinvent waste. A strong foundation is being laid that is forming opportunities for the next generation of sustainable waste managers, as well as for new waste economies that can be improved upon every day. There is more opportunity than ever to continue innovating and building upon the existing foundation of stewardship and sustainability. Let's recognize some of these waves of hope. These ways of thinking can be implemented by good stewards of any business, government organization, or community.

Strategies, Systems and Processes

LEAN (5 Steps):

Some early and very foundational ideas around reducing waste come from the Japanese LEAN five-step process. The Japanese car manufacturer Toyota is known for creating the process and, in doing so, improving manufacturing in the mid-1900s. The process is broken down into five steps. Step 1 is to identify value; step 2 is to map the value stream; step 3 is to create flow; step 4 is to establish pull; and step 5 is to seek perfection. The ultimate goal of the LEAN five-step system and process is to create a more efficient flow in a manufacturing process and reduce wasted time and resources. The process helps create better lead times and more accurate products in higher volume for manufacturing. LEAN ideas can certainly be used within any organization to reduce wasted time and resources, which should ultimately reduce physical waste. Many organizations currently use LEAN thinking and concepts for business modeling and process improvement (Lean Enterprise Institute n.d.).

Figure 39—LEAN Five-Step Process

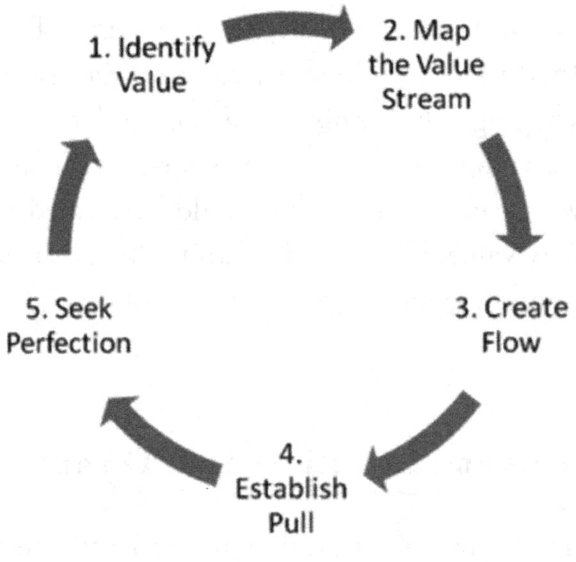

Source: The five-step thought process for guiding the implementation of lean techniques. Image copyright 2016, Lean Enterprise Institute, Inc. All rights reserved. Used with permission. https://www.lean.org/lexicon-terms/lean-thinking-and-practice.

Life Cycle Analysis (End-of-Life Plan for Materials):

According to the Global Development Research Center, life-cycle analysis (LCA) is a process of thinking that was adopted in the 1960 and 1970s. LCA is the concept of conducting a detailed examination of the life cycle of a product or a process and attempting to access the resource cost and environmental implications of different patterns of human behavior. LCA was developed to respond to and increase environmental awareness for the general public, industries, and government (Global Development Research Center 2018). Every product and process impacts the environment differently. For that reason, LCA can be used to account for all the inputs and outputs throughout the life cycle of a product or process. Whether it is the birth, design, raw material extraction, materials production, part production, or assembly, through to its use and final disposal, one can critically assess that point in the cycle. LCA is changing the way people and organizations are thinking about manufacturing and building, and this mindset helps those consider the environment when making business decisions. A related concept to LCA is design for environment (DfE), which is a design process that uses environmental impact assessment throughout the entire design process to keep impacts to zero or to an absolute minimum.

LCA and DfE strategies are making inroads in terms of waste reduction, emissions reduction, and implementing strategies to repurpose waste. The end-of-life phase in the life-cycle assessment is now, in some ways, the most important. Organizations are trying to come up with solutions for maintaining, repurposing, and creating environmentally friendly last-resort disposal methods, as well as economic opportunities.

Figure 40—Life Cycle Assessment   **Figure 41—LCA in Waste Management**

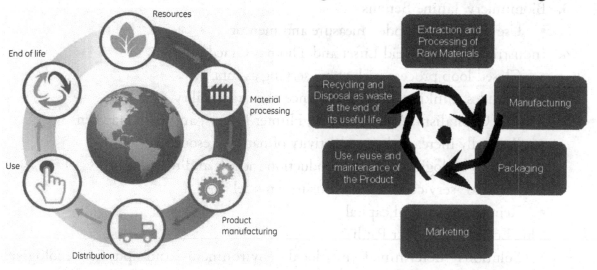

Source: © Copyright Cytiva, Reproduced with permission. Source: Global Development Research Center.

LCA and DfE thinking can be used for sustainable waste-management solutions. Cities and governments—and any organization, for that matter—can benefit from them regarding reducing waste or more properly and ethically handling waste. Many organizations, such as Levi's, use LCA for certain products and resources. If more companies and people adopt LCA thinking into their organizations, waste management will greatly improve. Thinking with the end in mind, first, can create a foundation for more sustainable thinking, ultimately helping the Earth and creating many economic benefits. Other models similar to LCA are Cradle-to-Grave analysis, eco-balancing, and material flow (Global Development Research Center 2018).

Circular Economy:

The Circular Economy (CE) idea and method was created and launched in 2010 by the Ellen MacArthur Foundation. Circular economy ideology has developed as a thought leadership method for many global businesses and governments, as well as for the academic world. The foundation focuses on five major areas: learning, business and government, insight and analysis, systems initiatives, and communications. The concept was spurred from various schools of thought, some dating back to the 1970s. These are:

1. Cradle to Cradle, William McDonough and Michael Braungart
   - Eliminate concept of waste
   - Power with renewable energy
   - Respect human and natural systems
2. Performance Economy, Walter Stahel
   - Closed-loop approach or functional service economy
3. Biomimicry, Janine Benyus
   - Use nature as a model, measure and mentor
4. Industrial Ecology, Reid Lifset and Thomas Graedel
   - Closed-loop process with waste serving as input
   - Includes thinking about the science of sustainability
5. Natural Capitalism, Amory Lovins, Hunter Lovins, and Paul Hawken
   - Radically increase the productivity of natural resources
   - Shift to biological inspired production models and materials
   - Move to "service-and-flow" business model
   - Reinvest in natural capital
6. Blue Economy, Gunter Pauli
   - Solutions determined by local environment and physical/ecological characteristics
   - Emphasis on gravity as primary energy
   - Hands-on focus
7. Regenerative Design, Ellen MacArthur Foundation
   - Foundation for circular economy framework

Each school of thought has helped lay the foundation for the circular economy approach. According to the Ellen MacArthur Foundation, a circular economy concept seeks to rebuild capital, whether this is financial, manufactured, human, social, or natural. This ensures enhanced flows of goods and services. The system diagram illustrates the continuous flow of technical and biological materials through the value circle (Ellen MacArthur Foundation n.d.).

The building blocks for circular economy are as follows:

1. Circular economy design
2. New business models
3. Reverse cycles
4. Enablers and favorable system conditions

Looking beyond the current take-make-dispose extractive industrial model, a circular economy aims to redefine growth, focusing on positive society-wide benefits. It entails gradually decoupling economic activity from the consumption of finite resources, and designing waste out of the system. Underpinned by a transition to renewable energy sources, the circular model builds economic, natural, and social capital. It is based on three principles:

• Design out waste and pollution
• Keep products and materials in use
• Regenerate natural systems

In a circular economy, economic activity builds and rebuilds overall system health. The concept recognizes the importance of the economy needing to work effectively at all scales—for large and small businesses, for organizations and individuals, globally and locally. Transitioning to a circular economy does not only amount to adjustments aimed at reducing the negative impacts of the linear economy. Rather, it represents a systemic shift that builds long-term resilience, generates business and economic opportunities, and provides environmental and societal benefits. The model distinguishes between technical and biological cycles. Consumption happens only in biological cycles, where food and biologically based materials (such as cotton or wood) are designed to feed back into the system through processes like composting and anaerobic digestion. These cycles regenerate living systems, such as soil, which provide renewable resources for the economy. Technical cycles recover and restore products, components, and materials through strategies like reuse, repair, remanufacture or (in the last resort) recycling.

The notion of circularity has deep historical and philosophical origins. The idea of feedback, of cycles in real-world systems, is ancient and has echoes in various schools of philosophy. It enjoyed a revival in industrialized countries after World War II when the advent of computer-based studies of nonlinear systems unambiguously revealed the complex, interrelated, and therefore unpredictable nature of the world we live in—more akin to a metabolism than a machine. With current advances, digital technology has the power to support the transition to a circular economy by radically increasing virtualization, dematerialization, transparency, and feedback-driven intelligence. The circular economy model synthesizes several major schools of thought. They include the functional service economy (performance economy) of Walter Stahel; the cradle-to-cradle design philosophy of William McDonough and Michael Braungart; biomimicry as articulated by Janine Benyus; the industrial ecology of Reid Lifset and Thomas Graedel; natural capitalism by Amory and Hunter Lovins and Paul Hawken; and the blue economy systems approach described by Gunter Pauli. (Ellen MacArthur Foundation n.d.)

Figure 42—Circular Economy Diagram

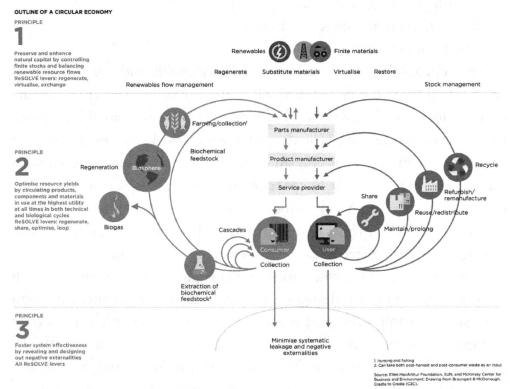

Sources: Ellen Macarthur Foundation, SUN, and McKinsey Center for Business and Environment; drawing from Braungart & McDonough, Cradle to Cradle (C2C).

The circular-economy approach might be the best strategy and systems-thinking approach for human activity yet. It puts nature (Earth) first. If people and organizations can adopt the circular-economy approach for all activity, it will reduce waste considerably. The circular approach should create new methods for reusing and reinventing waste, while promoting human development and activity that works harmoniously with the natural world. There are many companies and organizations following the circular-economy approach already, companies like Ecovative, Philips, TurnToo, and Caterpillar. These companies are using the circular-economy approach to create more sustainable business systems, which creates economic value, equality for workers, and environmental value all simultaneously.

The Natural Step:

The Natural Step (THS) framework and system was conceived by Karl-Henrik Robert in 1989. His inspiration came from the Brundtland Report that was published in 1987. The Natural Step framework and practice was conceived to help people and societies live more sustainably and in balance with Earth. This practice can be used specifically in sustainable waste management. The Natural Step is a foundation by which humans can live and create opportunities for individuals to reinvent waste.

The sustainability principles of The Natural Step are to live in balance with nature. The principles are broken down into four rules, which are:

In a sustainable society, nature is not subject to systematically increasing …

1. Concentrations of substances from Earth's crust (such as fossil CO_2, heavy metals, and minerals)
2. Concentrations of substances produced by society (such as antibiotics and endocrine disruption)
3. Degradation by physical means (such as deforestation and draining of groundwater tables)
4. In a sustainable society, there are no structural obstacles to people's health, influence, competence, impartiality, and meaning (The Natural Step n.d.)

Trademark of The Natural Step.

FOUR BASIC RULES DEFINE SUCCESS...
We call these rules 'Sustainability Principles'.

In a sustainable society, nature is not subject to systematically increasing...

1... concentrations of substances from the earth's crust (such as fossil CO2, heavy metals and minerals)

2... concentrations of substances produced by society (such as antibiotics and endocrine disruptors)

3... degradation by physical means (such as deforestation and draining of groundwater tables).

4. And in that society there are no structural obstacles to people's health, influence, competence, impartiality and meaning.

Source: The Natural Step Theory: https://thenaturalstep.org/approach/.

CSR, Triple Bottom Line, Global Reporting Initiative, ESG, and Others:

Over the years, the concept of corporate social responsibility (CSR) has evolved. In the mid-1900s, it was believed that companies and corporations existed to provide security for their workers, communities, and the United States. However, in the late 1900s, data and research showed that businesses and big corporations were negatively impacting the environment, on top of creating social inequality. Over the last twenty-five-plus years, more sustainable thinking in terms of putting people and the planet first have been implemented. A major sustainability framework was created in 1994 by John Elkington. Elkington coined the term and concept "triple bottom line—people, planet, profit" to create awareness and get businesses to account for and take inventory of environmental impacts and social impacts while maximizing financial gains (Clark 2017). The triple-bottom-line framework was greatly needed and was a significant step in the right direction.

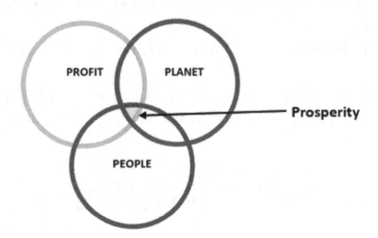

Source: Tyler Kanczuzewski, Inspiration from John Elkington.

Many organizations focus first on maximizing profit, without paying as much attention to their culture or quality of the planet. However, many organizations are starting to focus more on the well-being of people and the planet, just as much as—if not more than—making profits. After all, without people, or if the planet becomes uninhabitable, there will be no profits. Competition and capitalism are important, and they force organizations to constantly evolve and innovate. Green movements and sustainability initiatives are becoming more important to our US culture. They are forcing and encouraging organizations to implement sustainability, responsibility, and core values into their strategies. By implementing corporate social responsibly, triple bottom line, and green thinking, organizations can operate better for people and the planet. These strategies of green thinking can significantly reduce waste and create innovative strategies to repurpose waste. Balancing social, environmental, and economic health is a win for all.

Annual sustainability reporting became important for organizations big and small, for legal reasons, as well as to showcase care and depth of commitment to the planet. The Global Reporting Initiative (GRI) started in 1997 and has become a popular framework for many. Before GRI, environmental, social, and governance (ESG) existed to give organizations a balanced approach toward holistic sustainability, and ESG thinking is still widely used by many to report on sustainability.

Certifications and Standards

TRUE, Zero Waste Certification System (for Facility/Building):

Source: Courtesy of TRUE Certification for Zero Waste, and https://true.gbci.org/.

A new system called TRUE was implemented by the US Zero Waste Business Council (ZWBC) in 2013. TRUE stands for "total resource use and efficiency," and its aim is to help people and businesses better use and value waste. It is the first zero-waste rating system for buildings, facilities, and operations. TRUE uses the zero-waste definition created by the Zero Waste International Alliance (ZWIA) in 2002 to determine how to rate facilities based on waste reduction measures.

According to ZWIA, zero waste means a goal that is ethical, economical, efficient, and visionary, to guide people in changing their lifestyles and practices to emulate sustainable natural cycles, where all discarded materials are designed to become resources for others to use. Zero waste means designing and managing products and processes to systematically avoid and eliminate the volume and toxicity of waste and materials, conserve and recover all resources, and not burn or bury them. Implementing zero-waste principles will eliminate all discharges to land, water, or air that are a threat to planetary, human, animal, or plant health (Zero Waste International Alliance n.d.).

TRUE is now administered by Green Business Certification, Inc. (GBCI), which took over TRUE's management by ZWBC toward the end of 2016. As of 2018, over 304 million square feet of space was registered or certified with TRUE, and nearly 150 projects across thirty states and twelve countries were using TRUE (TRUE n.d.). The TRUE certification is for facilities, businesses, and good stewards who want to achieve zero waste through an accredited process and certification system that is internationally recognized.

> TRUE (Total Resource Use and Efficiency) Zero Waste certification is used by facilities to define, pursue and achieve their zero waste goals, cutting their carbon footprint and supporting public health. TRUE is administered by Green Business Certification Inc. (GBCI), the premier organization independently recognizing excellence in green business industry performance and practice globally. Established in 2008, GBCI is the only certification and credentialing body within the green business and sustainability industry to exclusively administer project certifications and professional credentials of LEED, PEER, WELL, SITES, GRESB, Parksmart and TRUE. GBCI is also a global certification provider for the EDGE certification system and the exclusive certification provider for all EDGE projects in India. Through rigorous certification and credentialing standards, GBCI drives adoption of green business practices, which fosters global competitiveness and enhances environmental performance and human health benefits. The goal of business, nonprofit and government facilities participating in the TRUE Zero Waste certification program is to divert all solid waste from the landfill, incineration (waste-to-energy) and the environment. Facilities achieve certification by meeting 7 minimum program requirements and attaining at least 31 points on the TRUE Zero Waste scorecard. TRUE is a whole systems approach aimed at changing how materials flow through society, resulting in no waste. TRUE encourages the redesign of resource life cycles so that all products are reused. TRUE

promotes processes that consider the entire lifecycle of products used within a facility. With TRUE, your facility can demonstrate to the world what you're doing to minimize your waste output. TRUE certified spaces are environmentally responsible, more resource efficient and help turn waste into savings and additional income streams. By closing the loop, they cut greenhouse gases, manage risk, reduce litter and pollution, reinvest resources locally, create jobs and add more value for their organization and community. TRUE certified facilities:

- **SAVE MONEY**: Waste is a sign of inefficiency and the reduction of waste reduces costs.
- **PROGRESS FASTER**: A zero waste strategy improves upon production processes and environmental prevention strategies, which can lead to larger, more innovative steps.
- **SUPPORT SUSTAINABILITY**: A zero waste strategy supports the three P's—people, planet and profit.
- **IMPROVE MATERIAL FLOWS**: A zero waste strategy uses fewer new raw materials and sends no waste materials to landfills, incineration (waste-to-energy) and the environment.

Implementing the requirements and credits of TRUE Zero Waste certification has the following benefits to any facility looking for a better approach to resource use and facility operations:

- Helps eliminate pollution—in our air, water and land—which threaten public health and ecosystems
- Improves their bottom line by reducing costs
- Cuts the ecological footprint by reducing materials, using recycled and more benign materials, and giving products longer lives by increasing reparability and ease of disassembly at end-of-life.
- Promotes positive forces for environmental and economic sustainability in the built environment by protecting the environment, reducing costs, driving the development of new markets, and producing jobs throughout our economy.
- Fosters strong total participation including training of all employees and zero waste relationships with vendors and customers
- Allows the facility to showcase their responsibility and commitment to the local and global community and the environment. (TRUE 2017)

Figure 43—TRUE Diversion

DIVERSION

All those seeking certification must provide waste diversion calculations. These numbers should be used consistently across all requirements and credits. The diversion rate should represent all activities within the project boundaries and include all materials generated within that boundary. Diversion should be calculated by weight as follows:

$$\text{Diversion Rate} = \frac{\text{Materials diverted from landfill, incineration (WTE), and the environment}}{\text{Total Generation}}$$

GBCI does not require a standard unit of measure. However, all materials must be tracked and calculated using the same unit of measure chosen. For more details on calculation of diversion, please see the Diversion credit section.

The following activities are all considered acceptable forms of diversion and may be included in the diversion calculations:

- **REDUCTION** – Efforts to reduce the generation of materials can be recognized in the diversion calculations provided that the reductions are documented from an established baseline representing previous operations

- **REUSE** – Avoided disposal resulting from the reuse of items

- **COMPOSTING** – Organic matter decomposed by micro-organisms into a soil amendment

- **RECYCLING** – Materials converted into manufacturing feedstock material and used in creation of new products (excludes use as fuel substitute or for energy production)

- **ANAEROBIC DIGESTION** – Organic matter broken down by microorganisms into a soil amendment in the absence of oxygen (byproducts must be recovered for productive use in nature)

- **OTHER PROCESSING TECHNOLOGIES**, not including incineration or waste-to-energy, in which the end product is recovered for productive use in nature or the economy

Source: Courtesy of TRUE certification for zero waste and https://true.gbci.org/.

Source: Courtesy of TRUE certification for zero waste and https://true.gbci.org/.

Leader in Energy and Environmental Design (LEED), Energy Star, Green Building Initiative, Living Building Challenge by International Living Future Institute (ILFI), Green Circle, and Green-e (Facility/Building):

- Facilities and companies can obtain the following certifications. All vary in degree regarding commitment and cost, but each enables facility and business owners to strive for more sustainable operations, from top to bottom, including reducing waste.

Green America Certified Business, Green Business Network (Business):

- Green America started in 1982 and has evaluated and certified over eight thousand small businesses as a Green Business. Green businesses adopt principles, policies, and practices that improve the quality of life for customers, employees, communities, and the planet. Members of the Green Business Network are changing the way America does business (Green America, n.d.). Green America certifies businesses that are using their businesses as a tool for positive social change, environmental sustainability and responsibility, are socially equitable and committed to extraordinary practices that benefit communities, and are accountable for continually improving and tracking progress in all facets of business (Green America, n.d.). Some companies that have been certified and supported through Green Business Network are Seventh Generation, Honest Tea, and Cliff Bar. Green America Certified Businesses and the Green Business Network is an opportunity for good stewards and companies to act more sustainably and find ways to better reinvent waste.

Certified B Corporation (B Corp) (Business, Corporation):

Source: B Lab Global.

In the mid- to late 2000s, the idea of establishing a newer and more innovative certification came to fruition. The certification is called B Corporation Certification (B Corp), which was developed and is operated by B Lab. Its premise was to create a certification process that allows companies to qualify by striving to use business as a force for good. According to B Lab, good means that businesses meet the highest standards of positive social and environmental, public transparency, and legal accountability to balance profit and purpose. According to B Lab, B Corps are accelerating a global culture shift to redefine success in business and build a more inclusive and sustainable economy (B Lab 2018). Large corporations like Patagonia and Cascade Engineering are Certified B Corporations. Both entities are committed to sustainable efforts and are making significant strides in physical waste reduction and eco-friendly material use. Patagonia makes a portion of its apparel line directly from recycled and/or organic material (like organic cotton). More than five thousand companies, from 155 industries and eight hundred countries, have their B Certification. But with over thirty million small businesses in the United States alone, there is an enormous opportunity for more companies to achieve B Corp Certification and for more good stewards to create B Corp Certified businesses and innovative strategies to reinvent waste.

Global Recycled Standard Process:

- The Global Recycled Standard (GRS) is intended to meet the needs of companies looking to verify the recycled content of their products (both finished and intermediate) and to verify responsible social, environmental, and chemical practices in their production. The objectives of the GRS are to define requirements to ensure accurate content claims and good working conditions, and that harmful environmental and chemical impacts are minimized. This includes companies involved in ginning, spinning, weaving and knitting, dyeing, printing and stitching in more than fifty countries (GRS 2018). This standard should help good stewards reinvent waste.

Life Cycle Assessment Certifications (Person, Process):

- The American Center for Life Cycle Assessment (ACLCA) is a nonprofit membership organization providing education, awareness, advocacy, and communications to build capacity and knowledge of environmental LCA. An ACLCA membership consists of industry, academia, government, consulting, and NGOs (ACLCA n.d.). The entity offers Life Cycle Assessment Certified Professional (LCACP), Certified Lifecycle Assessment Reviewer (CLAR), and Certified Lifecycle Executive (CLE) certifications. Life cycle assessment certifications encourage improved systems and processes for waste reduction and solutions for reinventing waste. Obtaining a certification is an opportunity for good stewards to sharpen one's skills and to educate others.

UL Certification (Product, Process):

- UL is an electrical safety rating system that is nationally recognized and accepted as the leading institution for electrical safety. UL has certifications related to sustainability and circularity (the concept of a product being created with its own end-of-life taken into account).

International Organization for Standardization (ISO) (Process):

- ISO was formed in 1947 in London. To date, the organization has published over twenty-two thousand international standards relating to technology and businesses, performing quality and efficient work. ISO creates documents that provide requirements, specifications, guidelines, and characteristics that can be used consistently to ensure that materials, products, processes, and services are fit for their purposes (ISO n.d.). Popular standards that ISO certifies are quality management (ISO 9001), environmental management system (ISO 14001), country codes, social responsibility, risk management, energy management, occupational health and safety, medical devices, and antibribery management systems (ISO n.d.). Many companies are ISO certified, or at least some of their divisions follow or adhere to ISO standards. However, could more companies strive for ISO certification? Also, could the International Organization for Standardization add new standards focused specifically on waste mitigation, stewardship, and sustainability?

Six Sigma (Person, Process, Organization):

- Six Sigma is a framework and practice used to improve business processes by using statistical analysis. The Six Sigma methodology is defined by the five DMAIC steps. DMAIC is a problem-solving initialism that stands for define, measure, analyze, improve, and control (Six Sigma n.d.). Six Sigma is specific training and is an approach based on data geared toward projects with quantifiable business outcomes. Through Six Sigma, processes are improved by controlling variation and understanding the intricacies within them, which results in more predictable and profitable business processes (Six Sigma, n.d.). The certification called 6Sigma.US is one of the main Six Sigma certifications. It assists organizations with all aspects of implementation processes from training champions, to certifying employees at various belt levels (Six Sigma n.d.). Champions are company executives and leaders who lead and implement Six Sigma projects and strategies to improve companies. There isn't currently much Six Sigma strategy and achievement specifically related to physical waste. But should there be, and would it help companies to reduce and better reinvent waste? One could imagine that champions could be certified to introduce waste-reduction strategies and save a company millions of dollars by using and reinventing scrap and other potentially wasted materials.

Carbon-Neutral Certifications and Initiatives (System, Process, Organization):

- Carbon Trust
- Carbon Zero
- Carbon Neutral
- National Carbon Offset Standard

Product Certifications for Stewardship and Sustainability:

- Cradle-to-cradle certified
- EcoLogo certified
- Green circle
- Green product certified
- Green seal certified
- Living product challenge
- Safer choice
- UL EcoLogo

Sustainable Material, Building Material, Product, or Packaging Certifications:

- Biodegradable certified
- Forest Stewardship Council
- Programme for the Endorsement of Forest Certification
- Rainforest Alliance
- Sustainable Forestry Initiative

Leave No Trace (Sustainable Material, Product or Packaging, and Principles):

© Leave No Trace: www.LNT.org

Fair Trade Certifications (Material, Product, Process or Organization):

- Cotton
- Dewing
- Factory
- Fair for life
- Ingredients

Agriculture Commitments or Certifications (Food and Agricultural Products):

- American Humane Association
- Animal welfare approved
- Bird friendly
- Certified humane raised and handled
- Certified naturally grown
- Farmed responsibly certified
- Marine Stewardship Council
- Monterey Bay Aquarium
- Ocean Wise
- Sea Choice
- USDA organic

Other Miscellaneous Commitments and Certifications for Potential Sustainability and Stewardship (Product, Organization):

- 1 percent For the Planet
- Clean Green Trsa Certified
- Epeat
- Fair Wear Foundation
- Global Organic Textile Foundation
- Green Pro certified
- Green Shield
- Made By
- Smart certified

All the above certifications, initiatives, and marketing logos are innovative solutions and are helping to improve stewardship and sustainability for products, materials, and packaging. However, we can still go above and beyond to create more aggressive methods to leave zero waste or obtain reduced-waste certifications and create initiatives and strategies to more efficiently value waste and reinvent it. There is great opportunity for individuals interested in creating and inventing!

Modern Innovations

Sustainable Business Centers, Eco-Parks, and Community Centers:

Sustainable business centers, eco-parks, business incubators, and community centers focused around waste are becoming a realistic concept across the country, promoting sustainability and cultivating education and innovation. Sustainability parks create environments for peers to learn from each other and to continually improve their practices. These parks create cultural mindset shifts and spur more holistic thinking in terms of reinventing and revaluing waste. A new business park concept has recently been created in Kent County, Michigan. This project was conceived partially to help the Kent County Department of Public Works reach its ambitious goal of reducing 90 percent of its waste (MSW) from going to the landfill by 2030. To achieve the goal, something innovative was needed.

A Briefing on the Kent County Business Park Master Plan:

Kent County has submitted a 106-page draft of the Department of Public Works (DPW) "sustainable business park" master plan, which estimated a $500-million direct private capital investment, with waste sorting and processing alone creating 150 jobs and an annual economic impact of $130 million. The new estimates come from 23 responses, representing 30 companies, to an RFI issued in March regarding the park master plan. The plan is to build a facility on 250 acres of land in Allegan County just south of the current South Kent Landfill. The park is meant to help the DPW reach its goal of reducing landfill waste by 20 percent by 2020 and 90 percent by 2030. County residents and businesses currently recycle about 8 percent of their waste. Of the nearly 500,000 tons of yearly waste in Kent County, an estimated 75 percent could be reused. In West Michigan, the estimated value of that discarded material —such as wood, metals, plastics and organics —is $56 million, and recovering and selling those valuables could create 370 jobs, according to a 2016 study by West Michigan Sustainable Business Forum. In Michigan, that equates to $386 million and more than 2,600 jobs. (Dawes 2018)

Source: Kent County Department of Public Works.

The idea and incentive for implementing the sustainable business park adjacent to the landfill is to create a more circular system in terms of waste management. The Kent County Department of Public Works wants all members and stakeholders of the county—and even those outside the county—to reimagine the value of trash. By doing so, people and organizations will hopefully better use waste and do everything possible to avoid sending materials to the landfill. For instance, the circular economy concept is a system that uses waste and finds ways to reinvent it. In the system, there may be some residual waste, but it is generally a much smaller residual volume compared to current waste-management systems. The new circular waste system is compared to a linear waste economy, where there is not a strong reuse loop of sustainable management of waste.

According to the Kent County Department of Public Works (Figure 44), the county processes over one billion tons of waste each year, and that number has been growing. Research shows that only 6–8 percent is being recycled and that roughly 75 percent of the waste could be reused, recycled, or converted into some type of energy. In Figure 44, 13 percent of the waste is plastic and 22 percent is paper. Both materials are extremely recyclable if solutions are enacted. This data shows how much opportunity there is to use waste, to see its value, and to find ways to reinvent it. Concepts like Kent County's sustainable business park make perfect sense and could be something that improves the entire waste-management system. Creating a new, sustainable waste-management economy that is much more circular and much better for our environment would be the ultimate game-changer.

Figure 44—Kent County Status of Waste and Proof of Opportunity

Source: Kent County Department of Public Works.

Source: Kent County Department of Public Works.

The proposed layout and location for the new business park (Figure 45) is just south of the current South Kent Landfill. This landfill will reach maximum capacity very soon, so either the neighboring property needs to be developed for landfill space, or another idea is required. That is where the sustainable business park comes in—to use the available land to create a new movement and opportunity for sustainable waste management ideas to be implemented. The proposed business park could generate many millions of dollars to the regional economy, as well as provide hundreds of new jobs. The park could lead peer-to-peer learning and create an environment that spurs innovation and creativity. The current economic impact assessments could actually underpin the real impact that may unfold. The business park is an exciting opportunity that could lead to big things, sustainably speaking, for Kent County, and should help reduce waste and achieve the ambitious goal of reducing 90 percent of waste from going to the landfill by 2030. This business park could lead to and promote sustainable waste-management movements and innovation hubs nationwide. It is electrifying for Michigan—and Kent County in particular—to embark on this journey of making a positive impact on the environment by reducing waste going to landfill, as well as by reducing waste that ends up across our precious land and waterways.

The infrastructure to build these sustainable parks might cost hundreds of millions of dollars in investment, and that is a significant obligation. However, the economic opportunities and positive environmental impacts over the long run will far surpass and outweigh the liabilities from an upfront investment.

Figure 45—Kent County Sustainable Business Park Preliminary Layout

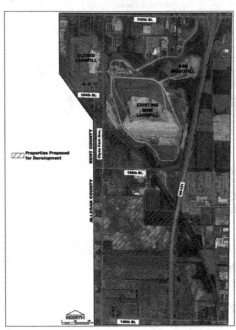

Instead of using these 200 acres for future landfill, we will develop a Sustainable Business Park that:

- Lays the **critical infrastructure** to support a regional circular economy
- Leverages **private sector development**
- **Attracts business** to localize the entire recycling or conversion process
- Preserves **open space**
- Expands **research**
- Generates and uses **renewable energy**
- Begins to **close the loop** in West Michigan

Source: Kent County Department of Public Works.

Three Important Nonprofits Reinventing Materials (Waste) Management:

The Product Stewardship Institute (PSI) creates a forum and system of ideas for producers of products, for consumers, for recyclers, and for governments to be product stewards, making products safer for the planet and people, from design to disposal. Simply put, PSI connects the public and private sectors to solve challenging municipal solid waste dilemmas. The Product Stewardship Institute embodies stewardship and sustainability for materials (waste) management. Concepts like extended producer responsibility (EPR) tie into PSI and overlap in many ways. EPR is a newer concept in the world of manufacturing and business. The idea is to essentially place more responsibility on the manufacturer in terms of sustainable end-of-life management plans of products or materials (resources). You can find their website at: https://www.productstewardship.us/.

GreenBlue is an environmental group focused primarily on sustainable usage of materials in society and the economy. It brings together a diverse group of stakeholders to facilitate and encourage sharing best practices and innovation to grow a more sustainable materials economy. The GreenBlue group is innovative and has launched a sustainable packaging coalition that encourages more environmentally friendly and sustainable packaging concepts for consumer products. You can find their website here: https://greenblue.org/.

The Recycling Partnership is a leading national group for improving recycling, focused on curbside recycling efforts and municipal solid waste. The partnership provides grants and funding, resources to improve practices, and partnerships for scalability. Thanks to nonprofits like this, reducing waste and growing recycling efforts to reinvent waste can be enhanced. You can find their website here: https://recyclingpartnership.org/.

TerraCycle—An Organization Enthusiastically Reinventing Waste

Founded by Tom Szaky and Jon Beyer in 2001, TerraCycle is potentially the most modern and innovative organization reinventing waste. The mission of the company is to eliminate the idea of waste and to recycle everything. According to TerraCycle, 202 million people among twenty-one countries have helped to recycle over 7.7 billion pounds of waste since 2001. TerraCycle supplies customers with zero-waste pouches, boxes, and pallets to collect hard-to-recycle items—like cigarette butts and tubes of toothpaste—to then return to TerraCycle. TerraCycle works with organizations to turn waste into new products or usable manufacturing materials. You can find their website here: https://www.terracycle. com/en-US/.

Loop is a new program launched by TerraCycle that implements a reusable and durable consumer products container, using reusable shipping totes and closing the loop in terms of packaging waste. Over thirty brands and 122 products in total are set up in the Loop system. So far, you can get groceries, beauty items, household essentials, health, and personal-care items through Loop. You can find their website here: https://loopstore.com/.

TerraCycle Global Foundation, a 501(c)3 nonprofit, is actively addressing the complex challenges of collecting and recycling waste in emerging countries and communities around the world. Some regions are developing quickly but have no infrastructure in place to handle the accumulating waste. The foundation offers education and supports collecting waste, with an emphasis on reducing plastic pollution in aquatic systems. More than $44.8 million has been raised to support these efforts since its 2001. You can find their website here: https://www.terracyclefoundation.org/.

Eco-Friendly Revolution:

Over the years, there has been a growing number of innovative companies and organizations using, producing, and supplying more eco-friendly resources and products. "Eco-friendly" at its root means that something is done or made without negatively impacting the environment. Some examples are using more renewable resources, recycled materials, and something that produces less waste. Eco-friendly products can decompose more quickly or can be biodegradable. Eco-friendly products can cost more than an alternative, less-eco-friendly product, but are people willing to spend more on a product that is better for the environment? Are people willing to invest more resources into eco-friendly and sustainable consumer products to better protect our natural world and reduce waste?

As discussed in step 1, respect, if humans adopt a sustainable mindset and put nature first, we can still sell products and goods that continue to drive economic opportunity, all while conserving our natural world. People, organizations, and companies are thinking more sustainably every day, but is it happening fast enough? There have been positive shifts in implementing more sustainability initiatives. Bulleted below are some examples of more prevalent or renewable resources in terms of the available biodiversity, as well as resources that are potentially safer, more eco-friendly, or recyclable. These materials and resources decompose faster, pollute less, and are all-around better for the natural world. Some of these materials are reused or recycled waste items, which ultimately help create circular economic systems (Bradley 2009).

- Aluminum bottles or cans
- Ashcrete
- Bamboo fiber
- Bamboo hardwood
- BPA-free plastic
- Biodegradable materials
- Clay brick
- Composite materials
- Compost
- Cork
- Corn starch biocompostables
- Corrugated cardboard
- Enviroboards
- Felt
- Grasscrete
- Hemp
- Low-VOC paint and finishes
- Natural earth clay and plaster
- Organic cotton
- Plantation-grown teak
- Plant-based products
- Plant-based compostables
- Permeable concrete or pavement
- Rammed earth
- Recycled or reclaimed goods and resources of many types:
  - Concrete
  - Foam
  - Glass
  - Metals
  - Paper and cardboard
  - Plastic
  - Polyester
  - Rubber
  - Wood
  - Wool
- Removed invasive plants and trees
- Reusable goods and products
- Rice hulls
- Sustainably sourced materials
- Soybean fabric
- Straw bales

Various Innovations and Inventions Starting to Reduce or Better Use Waste:

- Anaerobic digester technology, a form of waste-to-energy that converts organic waste (food waste) to contained methane. The contained methane can be used to create electricity. Approximately 22 percent (thirty-one million tons) of our landfills is food waste.
- Biogas technology to turn organic waste into cooking oil or fertilizer.
- Eco-friendly product companies have started focusing on making and selling eco-friendly goods, as well as moving toward sustainable operations.
- Efficient manufacturing and technologies that consume fewer resources, improve systems and processes, and help continually reduce waste.
- Large-scale composting commercially or independently. The compost can be used for community gardens or home gardens.
- New composite materials made from various types of recycled waste, such as plastics, foam, metals, and wood.
- New methods and systems for collecting litter and pollution to minimize health concerns and potential hazards for Earth.
- New recycling systems and technology that improve sorting and separation of recyclable waste in sorting facilities or material-recovery centers.
- Smart software and application technology to monitor or separate waste streams and collect reliable data on recycling streams.
- Sustainability think tanks for people and organizations teaming together to brainstorm new methods for collecting, repurposing, or reducing waste that would typically go to landfills.

Four Key Business Segments Reinventing Waste:

Below are four key business segments that are rapidly evolving and innovating to make more eco-friendly products and are starting to use waste as a resource. The companies listed per segment are not the only companies and organizations making a difference, but they provide great examples and are mentors for active stewardship.

Recycling Solutions, Systems, and Processes:

- Bata Plastics, Inc. (Plastics)
- Better World Books (Books)
- Feeding America, West Michigan (Food)
- Green Earth Electronics Recycling (Electronic recycling)
- PADNOS (Paper, plastics, metal, and electronics)
- Plastic Bank (Plastic trading)
- Schupan Recycling (Mixed recycling)
- TerraCycle (Mixed stream)
- West Michigan Compounding (Plastics)
- 4 Ocean

Consumer Products and Packaging:

- Allbirds
- Boxed Water Is Better
- Dixon Earth
- Dr. Bronner's
- Earth Choice
- EcoEnclose
- Etsy

- JUST Water
- Loop by TerraCycle
- Patagonia
- Seventh Generation
- Tentree
- United By Blue
- World Centric

New Technology and Engineering:

- Brightmark
- Grind2Energy
- HomeBiogas
- NatureWorks
- Precious Plastic
- Pure Energy Group
- SINTEF
- The Ocean Cleanup

Digital Platforms, Software, or Applications:

- Key Green Solutions
- MRM E-Cycling Management
- Recyclist
- Veoli

Organizations and Movements for Fighting or Reducing Waste Pollution:

Many organizations are fighting waste accumulation both on the front end and the back end. Front end means going to the source of waste creation, and back end means collecting waste pollution that is drifting around Earth over land and water. Back end is landfilling, recycling, and other modern waste-management solutions. There are organizations dedicated to collecting waste (e.g., plastic) pollution in the oceans, and some are actively coming up with solutions to make better use of waste. They are trying to reinvent waste.

- 1% for the Planet
- 4 Ocean
- Alliance for the Great Lakes
- Association of Plastic Recyclers
- Earth Day Network

- Global Plastics Alliance
- Global Recycling Network
- Great American Cleanup/Keep America Beautiful
- Greenpeace
- Institute of Scrap Recycling Industries
- Michigan Recycling Coalition
- National Waste & Recycling Association
- Nature Conservancy
- Ocean Conservancy
- Ocean Plastics Recovery
- Oceana
- Operation Clean Sweep
- Plastic Pollution Coalition
- Product Stewardship Institute
- Sierra Club
- Solid Waste Association of North America
- Sustainable Electronics Recycling International
- The Ocean Cleanup
- The Recycling Partnership
- Trash Free Seas Alliance
- West Michigan Sustainable Business Forum
- Wrap Recycling Action Program

Policy, Literature, Education, and Consumer Tips

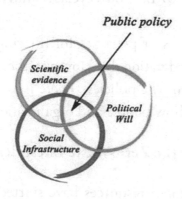

Policy Implementation:

To create significant change, policy tweaks or recommendations are necessary. If done correctly and fairly, policy implementation can nurture positivity. Some businesses and organizations are starting to ban single-use plastics, and some have banned products that are known to be detrimental to the environment. As shown in the picture above, it can take social infrastructure, scientific evidence, and political will to create public policy. The accumulation of waste over the years has started to create some sense of national urgency. The icing on the cake is that finding better ways to manage the waste creates new economic opportunities. Policy implementation can help advance a much-needed movement.

Examples:

- Zero-Waste Initiatives—Counties and cities are starting to create zero-waste initiatives by reducing the amount of waste going to the landfill by a certain percentage (Kent County Department of Public Works: 90 percent by 2030).
- Bans—Organizations, governments, states, and counties have banned the use of certain products because of negative environmental impacts (single-use plastics). Grocery store chain Kroger has vowed to ban single-use plastic bags by 2025. Kroger committed to ban the bags after admitting they use or supply one hundred billion plastic bags to customers annually.
- Pledges—Organizations and governments have pledged to reduce the use of certain products because of known environmental impacts or a lack of recyclability. For instance, McDonald's pledged to stop using Styrofoam cups in 1991, because Styrofoam was proven to be nonrecyclable and was negatively impacting the environment.
- Fines—Government fines or penalties for littering or dumping waste had been adopted and implemented nationally. However, there is still a lack of accountability, and there is ambiguity in the policies. There is an opportunity for much stricter enforcement regarding how people and organizations are getting rid of waste.

Literature and Education for the Zero-Waste Lifestyle:

Modern literature and educational resources have started to evolve and are addressing the importance of waste management and zero-waste lifestyles. Waste accumulation is a global epidemic. It obviously poses serious environmental threats but also wonderful opportunities for people and organizations to combat the epidemic head-on. By doing so, global movements will be created, spurring even more innovation and economic opportunities for all. Below are various books and educational literature that can help those wanting to learn more about waste-management solutions, as well as innovative ways to achieve zero waste. These writings may inspire more individuals to reinvent waste. Many documentaries are released regularly on the topic, so there is plenty of great information to dive into.

Books for People, Organizations and Policymakers:

- 101 Ways to Go Zero Waste
- Confessions of a Radical Industrialist
- Conscious Capitalism
- Cradle to Cradle
- Diet for a Hot Planet
- Garbology
- How to Reduce your Carbon Footprint
- Leave No Trace in the Outdoors
- Life Without Plastic
- No Impact Man
- Small Is Beautiful
- The Hunt for the Golden Mole
- The Upcycle: Beyond Sustainability
- The Zero Waste Lifestyle
- The Zero Waste Solution
- Zero Waste Home

Online Blogs, Videos, Educational Platforms, and Search Engines:

- Going Zero Waste
- Google or YouTube Search (Zero Waste)
- Litterless
- One Small Step by Now This
- Reddit | Zero Waste
- Sustainable Jungle
- The Zero Waste Memoirs
- Trash Is for Tossers
- Waste 360
- Wasteland Rebel by Shia
- Wild Minimalist
- UL Spot—Sustainable Company Search Engine

- Zero-Waste Chef
- Zero Waste Guy
- Zero Waste Home

Tips for Good Stewards: Ideas for Reinventing Waste

The following ideas are either new or inspired by various thought leaders. These ideas are not intellectual property; rather, they are intended for all individuals to adopt or to implement to reinvent waste and help make a positive impact on the planet. Some of these ideas and innovations have already been conceived in some shape or form by various thought leaders but have yet to be used on a broader national or global scale. Please note that none of these ideas have been tested or proven successful; they are potential solutions to combat the accumulation of waste and increasing pollution caused by waste. These ideas could specifically solve municipal solid waste, packaging, and plastic waste issues. These ideas could be used to create new organizations, companies, and new, sustainable waste-management economies. The ideas listed are not in hierarchical order and are part of four major categories: education, creating new policy, economic opportunity, and environmental activism.

Ten Ideas to Share Knowledge—for Conscious Consumers and Educating Stewards

- Ideas for all earthlings—"the producers of waste"—to reinvent waste and educate others how to improve. These stewards can help individuals reduce their footprints on the planet by being more conscious about their day-to-day decision-making and how it impacts the planet. Knowledge is power, and sharing is caring.

Ten Ideas to Help Implement Eco-Stewardship—for Environmental Stewards

- Environmental stewards can actively spread awareness of eco-conscious ideas that could reinvent waste and save the planet. We only have one Earth, so let's treat it better and not take it for granted. We can all strive to be eco-stewards.

Ten Ideas to Create Better Policies for Earth—for Policymaking Stewards

- Policy-focused stewards can promote change as well as creative ideas for government officials and policymakers to reduce our civilian waste, while encouraging consumers to reinvent waste. To really make a difference, we need better policies, locally, and globally.

Twenty-Five Ideas to Support Smarter Economies—for Business or Economical Stewards

- Economic and entrepreneurial stewards can create opportunities for utilities, commercial, industrial, and agriculture businesses, and organizations to implement and use the real value of waste. There is a market for reinventing waste now, and the value increases every day.

Ten Ideas for Conscious Consumers and Educating Stewards

1.	Educate all and courageously spread awareness about waste and general pollution.
2.	Produce less trash each day as an individual.
3.	Start your own movement and create positive change in the world.
4.	Support slowing or stabilizing the economy (consumerism) and grow sustainability.
5.	Cultivate an eco-conscious consumer mindset.
6.	Encourage fellow consumers to buy sustainable, eco-friendly, and quality products and packaging.
7.	Form or develop a new association, organization, networking group, or social club to promote zero-waste solutions and collaborative efforts.
8.	Campaign for organizations, especially restaurants, and people to adopt using and supplying reusable containers instead of throwaway disposables. Bring back the milk delivery program.
9.	Host zero-waste challenges or contests at work or at home.
10.	Create artwork made from waste to raise awareness about growing waste and pollution across Earth, as well as to show how waste can be much better used.

1. Educate all and courageously spread awareness about waste and pollution.

Landfilling, littering, pollution, the energy needed to recycle, and waste-management practices in general are all increasing rapidly. As citizens and earthlings, we need to educate each other more efficiently and swiftly about the growing need to reinvent waste, to reduce negative environmental impacts, and to create new green opportunities. Let's stop burying trash and polluting Earth, and instead help start a reinvent-waste revolution that realizes the value of waste.

2. Produce less trash each day individually and tell others to do the same.

If the average US citizen produces anywhere from four to seven pounds of trash per day, and trash is being buried in landfills every day, that means our country does not have enough systems and practices in place to sustainably manage all the new incoming trash.

Therefore, we must try to significantly reduce our individual waste to help level out the curve and allow sufficient time for new solutions and innovations to be implemented to help buck that trend. Waste is too valuable and too environmentally harmful, so it should not be buried in landfills anymore.

3. Start your own movement. Create positive change in the world!

Create a movement to raise awareness and get people thinking about waste differently in your local community. Waste should be highly valued as a usable resource. By reducing waste, we are potentially saving the planet and creating new economies that are genuinely more sustainable. The movement can be spread through organizations, various news media, social media, government agencies, and through word of mouth. The most important part of creating a movement is the delivery of factual data and having a mission or vision for individuals to combat potential issues head-on. Along with the delivery, clear steps are needed for good stewards and community leaders to implement to spread the word and continue a movement to reinvent waste. If the delivered message makes enough of an impression, word will continue to spread and create waves throughout local communities.

4. Support slowing or stabilizing the economy (consumerism) and grow sustainability.

As discussed earlier in step 1, respect, there has been a movement of overcoming logistical barriers that has spurred growth in capitalism, business competition, and consumerism. The movement has given birth to our modern-day globalized economic engine, an engine that provides seemingly unlimited available resources at the click of a button. As we now know, the problem is that we do not live in a world of unlimited resources. One solution for this serious problem might be to simply slow the economy down or just sustainably stabilize it. If we stabilize the economic engine, it may provide mass producers and companies time to implement more sustainable systems to reduce waste and increase efficiency. More organizations can strive to be more circular in their approach toward the use of resources and operations. If more organizations strived toward zero waste or large-scale waste reduction, the impacts would be immense. E.F. Schumacher wrote about the idea of social return on investment (SROI) in his book, *Small Is Beautiful*. The idea is to use SROI as a measurement tool for organizations and policymakers to incorporate social, environmental, and well-being returns into economic decision-making. One could argue that current economic decisions are not beneficial to the natural world and need to be adjusted to achieve more sustainable waste-management effectiveness.

5. Cultivate an eco-conscious consumer mindset and share that mindset with others.

Try to think more consciously when you are shopping for typical consumer goods. Buy fewer food and drinks packaged in materials that are hard to recycle. Buy fewer single-use plastics and disposable materials. Reuse anything you can, like a reusable water bottle. Buy products that are locally made, organic, or more naturally made, eco-friendly, manufactured using less material or more sustainable material, and are all around more eco-conscious. Grow your own eco-conscious mindset. Also, compost at your home or business and use the compost dirt at your home or local garden, or partner with an organization that will collect and use the compost.

6. Encourage fellow consumers to buy more sustainable, eco-friendly, quality, and longer-lasting materials, products, and packaging, even when costs can be higher.

There is a marketing term and demographic called LOHAS, which is an acronym that stands for lifestyle of health and sustainability. LOHAS is a segment of the market that believes in green initiatives, eco-friendliness, and sustainability. A relatively small population of people, compared to the overall population of consumers, are in the LOHAS segment.

It's a great time for more consumers to join the LOHAS movement and to buy products that are better for Earth. These products can cost slightly more than a comparable nonsustainably made product. However, as economies of scale and scope increase, the prices of sustainably made products will level out and potentially become cheaper than nonsustainably made products. Economic patterns can change, helping reduce waste and achieving higher sustainability. It starts with movements at the general population level, because the population (i.e., consumers) decide what companies and what products stay in business through their purchasing (i.e., votes). For instance, a beverage company will likely keep producing plastic bottles if consumers keep buying them. But if consumers stop buying plastic bottles and instead buy an eco-friendly alternative material, the movement will force that beverage company to produce an eco-friendlier bottle. Quality and longer-lasting eco-friendly products and packaging should be considered as well.

7. Form a new association, organization, networking group, or club to promote zero-waste solutions and collaboration.

One can always start a group that has the mission to brainstorm, promote, and encourage zero-waste solutions. An organization can be created legally, such as a formal 501(c)(3) nonprofit, or less informally through the creation of a social media page or general website. Either way, the goal is to encourage people to join the group and to collaborate, educate, and encourage innovation and to help make a difference in any community. These groups can be created locally, regionally, or even nationally. This goal is relatable to idea number three, to start a movement. The more people, the merrier; the more education, the better; and the more people uniting together can create a larger and more positive impact! The organization1 Percent for the Planet is a 501(c)(3) solely dedicated to raising money for environmental causes (as well as other nonprofits) across the globe and is making a huge impact. Creating organizations similar in nature to 1 Percent for the Planet, but that are more focused on ending waste and pollution, would be brilliant!

8. Campaign for organizations, especially restaurants, to adopt using and supplying reusable containers instead of throwaway disposables. Bring back the milk delivery service!

Campaign to bring milk delivery service back! Do you remember when a milkman came to routinely refill your glass milk bottles? Or when Coca-Cola refilled your glass bottles, or when reusable cloth diapers were a thing? Maybe that was before your time. But why did that change? People should campaign and encourage businesses, manufacturers, grocery stores, convenience stores, and especially restaurants to adopt using and supplying reusable containers instead of throwaway disposables. At a minimum, use packaging (plastic numbers/type) that are more recyclable or reusable, and sustainably made. Restaurants could create member programs for reusing food or drink containers. Some places are already doing this. This could also help reduce food waste.

9. Host zero-waste challenges, games, or contests at work or at home.

The big goal here is to treat trash with care, as valuable, and to find a home for it rather than sending it to the landfill. Try hosting zero-waste challenges at your home, in your community, or in your organization (gamification). Total volume saved or reduced during a certain period of time is a great option. Or you could focus a contest on particular material streams or total number of different materials collected and saved from going to the landfill. You can also provide rewards or incentives for the winners to spark their competitive spirits. Let the games begin!

10. Create artwork from waste to raise awareness about growing waste and pollution across Earth, as well as to show how waste can be better used.

A great way to showcase the potential value of waste is by turning it into artwork or something beautiful and creative. The aim is that it will inspire others to take waste and reinvent it into something else, as a valuable material and as something beautiful. There are so many ways to reuse or repurpose trash, and artwork using trash is the perfect way to display that. Let your creativity flow and make beautiful art from waste.

Ten Ideas for the Environmental Steward

1.	Clean up America campaign: **4R Earth.**
2.	Pick up one piece or one item of trash per day.
3.	Buy consumer goods/products from organizations that collect the packaging waste, start a reusability or recycling program. Join the reuse club!
4.	Create a "**Reinvent My Waste**" website, social media, and digital app platform that provides solutions for reusing waste. Also, find organizations that will accept waste to recycle or repurpose.
5.	Support campaigning and ending the era of single-use disposable packaging (mostly plastics).
6.	Start a "no more landfilling" global campaign and stop the throwaway lifestyle.
7.	Create a new website and blog platform to raise awareness for promoting and encouraging zero-waste lifestyles and societies.
8.	Develop and supply cleanup kits for litter and general pollution.
9.	Use or create a personal waste footprint calculator that includes tracking and weighing waste.
10.	Compost or create your own bio invention or innovation.

1. Clean up America campaign: 4R Earth.

Just like starting a movement and raising awareness, we can encourage the country to help in cleaning up America, just like the Keep America Beautiful campaign (or concept) and The Ocean Cleanup projects and organizations are doing to clean up litter and ocean waste. The group 4 Ocean is one that is cleaning up beaches all over the world, with an emphasis on collecting plastic. What we can do is clean up everything, from land to water, and to the air we breathe. We must keep all three of them clean and pollution free to sustain life on the planet. An idea could be to start a new campaign and movement called 4R Earth. New companies and concepts can be created by taking litter from our lands, waste from the ocean and waters, or toxic pollution from the air. People and organizations can reinvent waste to use it as a valuable resource while saving the planet. Who wouldn't want to save the planet while making a few bucks? At a minimum, all the waste collected through these types of campaigns should be in collaboration with recyclers or landfills to collect and more sustainably manage the waste, if there is no other existing solution for doing so.

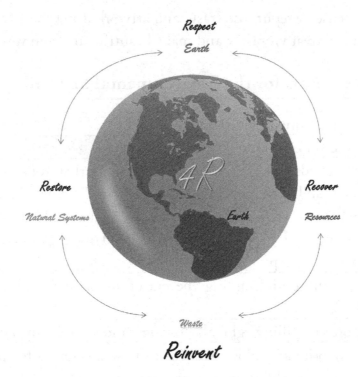

2. Pick up one piece or one item of trash per day.

The US population, as of May 2020, was almost 330 million, according to the United States Census Bureau. The world population is about 7.65 billion, according to the Bureau. Imagine if each person picked up just one piece of trash a day to reduce pollution and littering. The trash could be put in a trash receptor of some type or could be reinvented in some fashion. Let's estimate that the average weight of an item of trash is .5 pounds, meaning 165 million pounds of trash could be sustainably managed each day, nationally, and 3.83 billion pounds globally. That is a lot of trash, and that would be a lot of progress if a movement like this was to be ignited.

3. Buy consumer goods/products from organizations that collect the packaging waste and create a reusability or recycling program. Stop buying or refuse to buy products with packaging that create large volumes of waste and pollution. Join the reuse club.

This may be an extreme idea, but the fact is that many consumers and individuals are already resisting buying certain products, as well as certain packaging types. Create a member program or club for customers, allowing them to return the package (e.g., plastic) for reuse. A significant amount of plastic goes to our landfills and usually does not decompose for two-hundred-plus years. Millions of tons of plastic go into our oceans, waterways, and lands, breaking down into microplastics and creating potentially deadly toxicity to aquatic life. Consumers could stop buying plastic goods until more efficient recycling standards and practices are put in place by governments, organizations, and manufacturing businesses. Styrofoam is a product that is not efficiently recycled, or at least, not many systems have been implemented or invented yet to recycle and reinvent it. That opens the door for individuals who would like to figure out innovative methods for collecting and repurposing Styrofoam. Specifically, certain products could be completely banned or prohibited at businesses or governed regions. Some examples of this already include plastic bags, straws, plastic ring holders, single-use plastics, Styrofoam cups, and food not allowed to be in landfills. Many more items could be banned and prohibited, which should strongly encourage producers to make alternative and more eco-friendly products.

4. Create a "Reinvent My Waste" (or similar name) website, social media, and digital app platform and network that provides solutions for reusing waste and finding people or organizations that will take waste to recycle or repurpose.

This online platform for reinventing waste would include a pivotal search engine option where you can type in any kind of waste you would normally throw into the trash and get a list of ways to reuse it logistically or artistically at your home or business. You could also search for any nearby or regional organizations that will collect or accept your waste for repurposing or recycling. This concept is very similar to the How2Recycle.info label that helps individuals go online to find collectors or recyclers of certain material. If you find an organization that will collect it, you either drop it or ship it to that organization using eco-friendly methods. As for the printed informational literature, you would create general waste categories (plastic, metals, paper, wood, cardboard, glass, etc.) and then list organizations that collect those particular waste streams. The literature could be included in digital newspapers or published through its own magazine that is available to subscribers. The popularity of this "Reinvent My Waste" movement could create and encourage the innovation of new recycling organizations and facilities that could collect all types of waste for reinvention. This development should create innovative ideas for people and organizations to reuse, repurpose, and reinvent their own waste as well.

5. Support campaigning for and ending the era of single-use disposable packaging (mostly plastics). At a minimum, stabilize and create smarter consumption.

Single-use packaging is too common and probably too convenient. It has no doubt made life easier for most people. However, we have landfills near capacity across the globe and pollution is mounting, with estimates that food and beverage packaging comprise almost 50 percent of modern landfilled waste. Most recyclers do not have systems in place yet to successfully recycle much of the single-use packaging and consumer goods. One solution is to stop buying, or to greatly reduce consumption of single-use goods in packaging such as hard-to-recycle plastics (like grades #4, #7, and so on) and Styrofoam. Consumers and buyers of products and packaging can help make swift changes by not buying certain items and by campaigning for others to do the same. One could also campaign directly to the manufacturers and suppliers of the products and packaging to request that they end single-use packaging and switch to alternatives that are collectable, refillable, reusable, compostable, easily recyclable, or sustainably made.

6. Start a "no more landfilling" global campaign and stop the throwaway lifestyle.

Someone could get super assertive and forward thinking by proposing or campaigning to put an end to landfilling. Landfills are at or near capacity, land is scarcer than ever, and levels of trash and pollution are increasing, especially in developing countries. Microplastics are contaminating our precious food and water resources. What will it take for the Great Lakes region, the United States as a whole, and the world to decide that landfilling potentially needs to end? It may be an extreme idea, but it might be one of the only solutions we have. We also must campaign to end the throwaway lifestyle that many individuals practice, because most of the single-use disposable consumer goods packaging is very hard to recycle. All food waste that is being landfilled—that comprises about 24 percent of landfills—can easily be composted and used for agriculture.

7. Create a new website and blog platform to raise awareness for promoting and encouraging zero-waste lifestyles and societies.

The idea here is to make zero-waste challenges (e.g., zero waste for thirty days) and various fun activities and ideas for collecting waste to repurpose at a home or business (say, for instance, artwork). The platform could include educational data on waste accumulation and pollution. The website and blog should be user friendly and include a mobile application. The platform could have a social media or blog component to it to allow users to connect from all geographic regions. Users could encourage each other to be better stewards and to minimize individual waste and ecological footprints. Included could be a commerce and trade page for buying and selling specific materials (waste) from one another, like a trade-and-barter program. Zero-waste bloggers could write about picking up waste on the beach of Lake Michigan during a vacation. That person could give facts about how many pounds of waste were collected, the total items, types of material, or even some information about the brands or the manufacturers of the waste.

8. Develop and supply cleanup kits for litter and general pollution.

This concept is similar to dog-waste bags and containers you would find attached to a trash can or light pole on a sidewalk of a city park. Instead of a small plastic bag for dog waste, you would need to provide a bigger or more heavy-duty bag, potentially with safety gloves. You could stock these kits in safety containers at beaches, roadways, waterways, city and state parks, and other communal areas. The trash that is collected could be sent to the nearest recycling center or a dumpster for worst-case scenario. Another option is to ship the collected waste to the cleanup kit company or to deliver it to the nearest location. The cleanup kit company could then use the trash to make new products for sale or categorize the waste streams and try to sell or supply the material to various vendors. This organization could be both for-profit or nonprofit and rely on funding from sponsorships, donations, or business contracts.

9. Use or create a personal footprint calculator for tracking and weighing waste.

There are plenty of phone applications and programs to help individuals track personal body weight loss or other goals. Surprisingly, there are almost no programs for tracking how much trash people actually throw out on a daily basis, what type of trash is thrown out by percentage, and what individuals' human footprints are from that daily waste. National tracking federally (through the EPA) and statewide governments track recycling and landfilling to calculate total waste and breakdown by type. A personal footprint concept could easily be conceived on a cell phone, tablet, notebook, or on a computer. Pitch this idea to an app developer and see if they are willing to draft up a program or design! Make sure to buy a scale so that you can start weighing your waste!

10. Compost at home, at your business or organization, or at a community center. Better yet, create your own bio-innovations and inventions.

In 2018, thirty-five million tons of food waste went to landfills, totaling 24 percent of all waste landfilled. That made up the single largest stream being landfilled, which means it could be the most underutilized as well. So, what can be done on a broader scale? For starters, people can compost the organic food waste at home and use the homegrown compost soil in their own gardens. Individuals can also compost organics at their businesses, organizations, or community centers, and use the compost for a community garden or sell or supply it to organizations that consume compost. Homegrown organic compost is high in nutrition and quality and can be well utilized in any garden to cultivate new fruits and vegetables. Another idea is to take food waste and utilize technology to turn it into a biogas or another type of energy. There are plenty of bio-innovations or new inventions to be created! Start a bio club in your town to cultivate new ideas for reinventing food waste, as it could help reduce a significant amount of the waste filling up our landfills.

Ten Ideas for the Policymaking Steward

1.	Implement minimum tipping fees at all landfills.
2.	Ban or heavily limit disposable (single-use) packaging (focusing on plastics).
3.	Implement zero-waste policies, plans, and incentives.
4.	Improve recycling infrastructure. Push or enforce recycling companies to collect more materials.
5.	Create waste-credits programs, similar to carbon or renewable energy credits.
6.	Kickstart an eco-friendly and stewardship certification for products and packaging (focus on reduced resource use, recyclability/compost ability, and reusability/reinvent-ability).
7.	Grow and improve third-party certification and testing of eco-friendly products and packaging.
8.	Create state- or national-level programs to penalize manufacturers (producers) and organizations for littering and general pollution and volume of waste landfilled. Make producers more responsible!
9.	Campaign for federal increases in littering and pollution fines/citations for all citizens.
10.	Form new EPA mandates and regulations per state on permittable amounts of waste that can be landfilled, as well as improved emitted methane monitoring and control.

1. Implement minimum tipping fees at all landfills.

A tipping fee is how much a person or organization is charged per tonnage of municipal solid waste dumped or delivered to a particular landfill. As mentioned earlier, tipping rates vary on region, but the average tipping rate in the United States in 2015 was around forty-eight dollars. Imagine if the average tipping-fee rate was to increase significantly. People and organizations would likely try more eco-friendly means of disposing of waste and increase recycling efforts. It would be important to make recycling systems and infrastructure more affordable than landfilling. The net economic benefit of improved recycling versus a declining landfilling industry would likely be net positive. Some cities on the East Coast have already had to increase average tipping fee rates due to space restrictions. If communities and organizations cannot come to a consensus on increased rate structures, government institutions may need to step in to help start the process or make recommendations.

2. Ban or heavily limit disposable (single-use) packaging (focusing on plastics).

In 1955, *Life* magazine celebrated "throwaway living" to adopt the mindset that single-use plastics were convenient and brilliant. Plastic innovation exploded during the 1950s' space race. That was the start of widespread usage of single-use, disposable plastics and products. Such usage is still going strong today and might even be growing in some respects. What if a movement was created to ban or heavily limit single-use, disposable products? The movement could propose reusability or recyclability as the new standard for consumer products and the packaging they come in or are shipped in. This movement to ban and limit single-use, disposable products would encourage buyers to buy reusable or recyclable goods and encourage manufacturers to produce reusable and recyclable goods. MSW could be reduced by nearly 25 percent if single-use products and disposable products are not sent to the landfill. Pollution would also be greatly reduced.

3. Implement zero waste policies, plans, and incentives.

More cities and states can strive to go zero waste. Many cities and states are currently trying to hit renewable energy goals, and going zero waste is similar. Goals like reducing 50 percent of total MSW going to the landfill by 2025 could be adopted for many cities and states. Some newly established towns and eco-villages are committed to aggressive sustainability standards and measures, such as being 100 percent powered by renewables. These eco-villages and cities could commit to zero-waste standards and policies. More city and state fines and penalties should also be put in place for littering and pollution. Many governments already impose fines for littering, but still more local and national government institutions could adopt such penalties, or at least increase them. In certain regions, indoor/outdoor composting could be enforced (some countries in Asia already do this). If you don't have use for the dirt created from the compost at your own home or garden, you can take the compost to a local or community garden. The United States could also create federal initiatives or mandates to minimize landfilling as a method for handling waste. Instead, waste could be recycled or reinvented nationwide to better use the real value of waste, consume fewer natural resources, and create more biodiversity, which is needed to create to products and goods.

4. Improve recycling infrastructure. Push or enforce recycling companies to collect more materials and raise the value of recyclables.

Since the 1970s, consumers have been reimbursed for physically dropping off recyclable goods at recycling centers (aluminum cans, glass bottles, etc.). Reimbursements are typically five to ten cents per piece, depending on the state, but not all states offer it.

Curbside recycling started in the 1970s, where consumers or end-users could pay (monthly or yearly prices) for a recycling container and pickup service. New potential solutions are having recycling centers pay end-users more for deposited items, charging less for recycling services, or adding more material streams to the acceptance list. Reimbursements could increase by 100 percent on many materials if real, potential values were assessed properly and proper infrastructures were put in place to efficiently reinvent waste. On top of that, "free" recycling services, such as a bin to store and a recurring pickup service, should be implemented. As waste streams start becoming more valuable, recycling service centers would make more money through vendor contracts. Then they would not need to rely on service fees upfront to cover overhead and revenue needs.

5. Create waste credits programs, similar to carbon or renewable energy credits.

Just as with energy credits or carbon credits, the United States could implement a waste credits program. The idea here is to set thresholds for how much waste individual households and organizations can send to the landfill. If you waste more than your allotment, you would have to buy credits from certified trading institutions. Conversely, if you are not sending much waste to the landfill and are accumulating credits, you could sell the credits for revenue. A waste-credits program would create a brand-new market that could make a significant positive impact on the country and on our waste crisis. The waste credits would be monitored and transacted through a trade market, mostly through third-party firms. This idea correlates with number ten and would especially help reduce methane emissions.

6. Kick-start a stewardship certification for products and packaging. Focus on reduced energy or resources, recyclability, compostability, reusability, and reinvent-ability.

There are no national certification standards specifically for eco-friendly products or packaging. Across the United States, there are certifications like Fair Trade, Energy Star, Green Seal, and USDA Organic. These certifications are more directly related to sustainability in general. There are currently no certifications that are directly related to minimizing waste and maximizing efficiency. There is the new TRUE zero waste rating system by the GBCI, but that pertains more to a complete business system and operational waste. It is time for a certification and labeling system specifically for individual products and packages. The certification can base its focus and ratings on three main areas: minimal resource or smart-energy use, eco-friendly/compostable material use, and reusability/recyclability/reinvent-ability. The rating system and certification can be a combined total based on the number of resources used or total footprint to make, total volume of waste produced, and a reusability/reinvent-ability factor.

7. Grow or improve third-party certifications of eco-friendly products or packaging.

To decrease confusion on what is eco-friendly and what is not, there could be general improvements in certification practices as well as in organizations. "Eco-friendly" can mean less material or energy went into the manufacturing, renewable energy was used, more renewable or quickly regenerative resources and materials were used, or more recyclable material and/or compostable materials were used. There should be some type of scoring or rating system to compare the true sustainability of different products and their packaging. This rating system (referencing idea number six), would help ease confusion in understanding the level of eco-friendliness and overall sustainability, as well as to allow good stewards to make more eco-conscious decisions every day. The FSC, as well as others, promote sustainability and eco-friendly practices and can be modeled upon.

8. Create state- or national-level programs to fine or penalize manufacturers (producers) for littering and general pollution, or volume of waste landfilled. The goal is to improve circularity, and put more responsibility on producers, like the EPR methodology.

Would it be too extreme to penalize the manufacturers of products and packaging for their consumers' waste that is landfilled in extreme amounts or found in waterways and land? It might be time to look at creating a national program, or state-level programs, to hold manufacturers more accountable for the waste created by them and their consumers. EPR is a framework concept and idea gaining national attention, with a goal of putting more attention and accountability on manufacturers in terms of end-of-life planning and waste mitigation. If the situation continues deteriorating, the creation of specific laws or policies might be necessary to enact.

9. Formalize federal increases for littering and pollution fines/citations for all citizens.

To create a sense of urgency and spread awareness across the country, a federal mandate could be put in place to raise fines or citations for any offender of littering or general waste pollution. Some states are already stricter on trash-related policies than others, but states that are not enforcing these policies could be urged to strengthen their policies. Pollution is a serious epidemic and is growing worse globally. People should not only think about a healthier Earth but also about avoiding citations.

10. Form new EPA mandates and regulations per state on permittable amounts of waste that can be landfilled, as well as improved emitted methane monitoring and control.

This idea correlates to number five; however, the main difference here is to get the EPA much more involved with regulating the amount of waste we can landfill as well as the amount of methane or other GHG emissions that can be emitted nationally. A good way to break it down is by state and population (per capita). It's an important time for the federal government to step in.

25 Ideas for the Entrepreneurial or Economical Steward

1. Create circularity methods and programs for product and packaging material.
2. Form Operation Green Sweep (OGS) for better stewardship and sustainable practices.
3. Develop or build recycling hubs at every landfill in the United States.
4. Build salvage (junk) yards for all waste streams.
5. Improve technology, sorting, and separation systems and processes for all recycling centers.
6. Reinvent landfill efficiency and design and consider landfilling only as a last resort for waste.
7. All manufacturers should develop a product end-of-life cycle plan or implement detailed recycle guidelines for every product or good manufactured.
8. Produce user-friendly waste scales for homes, businesses, and organizations.
9. Form dollar-per-pound charging programs for trash collection firms, organizations, and landfills.
10. Tax incentivizing or federal rebate programs for those who go above and beyond managing waste.
11. Start zero-waste venues and events companies (e.g., airports or travel centers, stadiums, concerts, sporting events, business conferences, and various other communal sessions).
12. Build or create new products, buildings, and technologies from reinvented waste.
13. Create eBay-, Craigslist- or Facebook Marketplace–style business platforms for waste streams to resell and trade with others, something similar to freecycle.org.
14. Mine landfills as well as new and creative solutions to collect waste from all places.
15. Start zero-waste, sustainable, or eco-farms, shopping centers, supply stores, grocery stores, restaurants, manufacturers, transportation firms, logistic firms, businesses, or other eco-organizations.

16. Create a B-certified corporation or start an entrepreneurial career focused on reinventing waste.
17. Build more plastic banks or waste banks, especially in populated or heavily trafficked areas.
18. Develop and invent new eco-friendly or more innovative plastics.
19. Procure sustainably farmed, sourced, and eco-friendly raw material in the manufacturing of products.
20. Use more plant-based materials, compostable packaging, and bio-solutions.
21. Increase waste-to-fuel, energy, or new material innovation.
22. Start waste or litter cleanup (recycling) and maintenance companies. Examples could include hiring a firm to clean up sections of highways, waterways, cities, or beaches.
23. Build eco-parks or sustainability parks, sustainability think-tanks or incubator hubs, community centers, and recreational areas.
24. Create a composting company to build innovative products and systems for better composting.
25. Produce great and long-lasting quality products that are made with eco-friendly materials. Include free or low-cost repairs and returns as a pivotal part of the business platform.

1. Create circularity methods and programs for product and packaging material.

Whether you are a producer of packaging materials or of products that need packaging for safety and logistical reasons, there is an opportunity for you to create circularity. Reusable, durable, and sustainable material is usually the optimal solution. For instance, if you ship food or other items in cardboard boxes from point A to B, the recipient could empty the boxes and send them back to you on the next route run. One could also use more wood, cardboard, metal, or durable/reusable plastics for packaging. For convenience food and beverage producers, it's hard to package the food or beverages in reusable packaging that is cost effective. However, you could have consumers collect the packaging or wrapping and send it back (like the firm TerraCycle) to recycle or repurpose it, or instead, you could reinvent packaging. Loop by TerraCycle is piloting this concept with select consumer-brand products, where the food container or product container is reusable and returnable, eliminating the product and packaging waste. There are great opportunities for producers to create more circular programs for all types of products and packaging and to make more sustainable closed-loop systems. The design concept DfE is one you can model after as well, as the intent is to make something that is in harmony with nature.

2. Form OGS for better stewardship and sustainable practices.

Every organization or business can run an OGS. OGS is a new concept for creating strategic sustainability or "green teams" within organizations to help implement good stewardship and sustainable practices across the organization, at all levels. The idea is to sweep across the organization to look for and research potential problems with systems and processes, and to then implement solutions for more positive financial gains, social equity, and measurable environmental impacts. The task force should involve people from each division or area of the company, and at all levels of hierarchy. Team members could be rotated off and on. These types of teams could grow any organization significantly, if done efficiently. The idea comes from the Operation Clean Sweep concept, a certification and commitment in the plastics industry for producers to actively reduce plastic waste to an absolute minimum, and especially out of the environment. The most important part of Operation Clean Sweep is the auditing to confirm minimal waste leaving the building or entering the environment. So, OGS would similarly audit the organization regularly for sustainability improvements environmentally, financially, and socially.

3. Develop or build recycling hubs or incubators at every landfill in the United States.

This idea comes from the Kent County Department of Public Works proposal for developing a sustainable business park next to the South Kent Landfill. The goal of the proposal is to promote recyclers and companies to set up operations at the park to collect waste as a resource, thus diverting waste from going to the landfill. Implementing a national campaign for installing recycling hubs or incubators at every landfill location would be a difficult and expensive task. There would have to be an incentive for landfill owners to invest in a recycling center, and the recycling center would have to be efficient for vendors to buy or take certain materials from the recycled waste. To ensure the potential success of this idea, a federal policy or incentive may need to be adopted to guarantee some accomplishment. Imagine the impact in waste reduction and utilization if a recycling center were installed at each landfill. The idea would be for waste to go through the recycling center or hub first, hopefully finding a home instead of being sent to the landfill. The opportunity could generate new economic opportunities. Imagine an industrial drive-through area where you could drop off a truckload of a specific material stream to the highest bidder.

4. Build salvage (junk) yards for all waste streams.

Across the United States, there are thousands of junkyards owned and operated as businesses. Most of them are old vehicle graveyards. Better than sending them to a landfill, right? At these salvage centers, you can buy or sell items very economically. So, the idea here is to create junkyard-style business centers for select material streams (plastics, glass, wood, metals, and so on). Some items may need to be stored under a roof or containerized in some way to keep them dry or from blowing away. Cheap land is out there and available, and rundown buildings exist, and they could make great locations for storing and selling salvageable material.

5. Improve technology, sorting, separation systems, and processes for recycling centers.

Improve recycling center equipment technology. Sorting and separating efficiency can greatly improve, such that specific material streams are better separated, cleaner, and more organized. The cleaner and better organized materials are, the more valuable and desired they are. The recycling industry has already been making big strides in this direction; however, there are still significant opportunities for improved efficiency and better systems and processes. Finding a balance between machine and human touch will play a significant role in building state-of-the-art recycling systems and processes to maximize the amount of recyclable, recoverable, and reinventable materials. Consider even drive-through areas for customers to dump waste streams in strategic areas, making it easier for both people and businesses to recycle.

6. Improve landfill efficiency and design and consider landfilling as a last resort option.

Improve landfill efficiency and design and continue to innovate and improve processes. Ensuring environmental safety for every landfill is extremely important. Improved methane capturing rates and efficiency to better use methane-to-energy is critical, as well as protecting the environment from any leaked methane. (Methane is one of the worst greenhouse gases that negatively impacts the atmosphere.) Furthermore, landfills should be considered as a last resort for waste that has no use or no potential value. The need for innovation opens up opportunities for those with entrepreneurial spirit. If we are going to produce methane, we need to manage it safely and more efficiently.

7. All manufacturers should develop a product end-of-life cycle plan or detailed recycle guidelines for every product, material, or good manufactured.

Product, material, or company managers can consider (especially before producing and supplying a new product to the market) determining a strategy for an end-of-life process or program for all products, materials, and goods. They should strive only to manufacture products and materials that are reusable and/or that have no end of life. If the product must have an end of life, it should be biodegradable, recyclable, or compostable. Another potential sustainable idea is for product, material, and goods producers to make products and packaging that last much longer or that are reusable. That way, consumers don't have to buy excessive amounts of the same identical product.

8. Produce user-friendly waste scales for homes, businesses, and organizations.

Imagine if dumpsters or waste bins had scales or if waste and recycling collection trucks had scales to weigh your trash or recyclables. The idea here is that weighing waste happens on the front end so that people or companies can keep more accurate counts of their total annual waste. This could be a great business opportunity if a viable solution is created. Industrial scales already exist, so this should be a viable solution. If people knew how much they threw out daily, weekly, monthly, or annually, they would likely reduce the amount of trash they are throwing out.

9. Form dollar-per-pound charges from trash collection firms, organizations, and landfills.

Rolling off idea number eight, if accurate accounting of waste took place, landfills, trash-collection firms, and various organizations could charge customers based on pounds generated rather than general service fees. This concept could be more lucrative for the trash collectors in the short run and help encourage trash creators to reduce their waste in the long run, thus helping reduce negative environmental impacts. Sometimes, the best business ideas are ones that will make your business obsolete in the long run, but if you continually innovate and create new products or ideas, your company should endure.

10. Tax incentivizing or federal rebate programs for those who go above and beyond in managing waste and pollution.

This idea is a blend of policy creation and entrepreneurial incentive, both for the policy and entrepreneurial waste steward. People or organizations should be recognized and rewarded for their efforts to reduce waste and pollution, for the health of our environment, and for all people. For this to happen, the local, state, and federal governments would have to implement new standards (incentives or rebates), structures, and policies for individuals pursuing waste and pollution management. Governmental support could slingshot the waste- and pollution-management industry for the betterment of society.

11. Start zero-waste venues and events company (e.g., airports or travel centers, stadiums, concerts, sporting events, business conferences, and various other communal sessions).

Big events produce a lot of waste because a lot of people are packed into one area, usually consuming large amounts of food and beverages, paper products, and souvenirs. These types of events typically sell or supply food and beverages in disposable containers, usually being plastic that is likely not eco-friendly. These host companies and people in leadership roles could decide to promote reduced waste, recycling focuses, or complete zero-waste events. Eco-friendly and biodegradable products and containers could be used to achieve zero waste, or even reusable containers that can be returned, cleaned, and used at a following event. Achieving zero waste for large events would be a significant accomplishment and provide encouragement and hope to many that zero waste is, in fact, achievable. This idea is for existing event centers that can evolve to zero-waste initiatives or for new centers to build their missions around zero-waste concepts.

12. Build, invent, or create new products, packaging, building materials, or equipment from reinvented waste.

There are endless possibilities and opportunities to use waste as a valuable resource. The book *The Upcycle*, by William McDonough and Michael Braungart (authors of *Cradle to Cradle*), coined the term *upcycle* to demonstrate that life and waste can be reimagined, and that humans can live more harmoniously and sustainably with nature. Another idea is to make things and use materials that last much longer, or forever, and that concept is designing for abundance. Rather than downcycling things to a landfill, the upcycling method proposes the possibilities of giving materials a chance for a new life or for a better life. The book even proposes better methods for managing water and sewer waste. Bouncing off the idea of upcycling, one could take different material streams and merge them to form new composites. Individuals could even invent new composite materials (from waste streams) and use them for building or developing cutting-edge products, building materials, consumer goods and other innovations.

There are many good stewards and organizations that are already developing new composite materials and leading initiatives in that field of work. There are great opportunities for those individuals who want to follow this same path and create brand-new inventions and innovations. The best part is that most of the raw material (trash) is free. You may have to figure out a way to collect it, sort it, and clean it to get a usable feedstock, but the time and resources needed should not outweigh the fact that the material is free on the front end. There is a company called MacRebur that is making roadways and pathways out of plastic. Patagonia can put recycled polyester and plastic into its new manufactured apparel. These are the types of solutions we need more of, so that society can holistically reinvent waste.

13. Create eBay-, Craigslist-, or Facebook Marketplace–style business platforms for waste streams to resell and trade with others. Something similar to freecycle.org.

Imagine an online business platform for buying, selling, or trading waste. It may seem strange or unlikely, but it could be a great opportunity. This would open up an entire new economy that could reinvent waste to great extents. In some ways, it would be like an online junkyard; however, there would be more organization in terms of material-stream inventory and pricing. There is opportunity for end-users, regular businesses and organizations, manufacturers, distributors, digital companies, and logistics firms to all get involved. One thing that will need to be considered is the carbon footprint of transporting waste. It is certainly eco-friendly to divert waste from going to the landfill and to find ways to repurpose it. However, if the carbon footprint to transport the waste over long distances is high or threatening to the environment, alternative solutions for managing the waste should be considered. Smart and efficient eco-friendly transportation methods should be considered for logistics and for shipping of all forms of waste. Materials Marketplace and Scrapo are both online hubs that are actively operating in this fashion, where people or groups list certain material streams and specific amounts that they are willing to sell. Listing for free is possible, too, where one can have the buyer pay for shipping. The Freecycle Network offers a website (freecycle.org) where people can list items to give to others for free, versus not wasting them. Freecycle.org is a great platform to potentially model after.

14. Mine landfills, as well as find new and creative solutions to collect trash from all places.

According to landfill safety, regulation, and best practices, most landfills (or cell sections) can be reopened or accessed about thirty years after being completely sealed. There are many landfills (or sections) that have now been sealed for thirty-plus years, and many more landfills that will hit thirty years very soon. That means there will be billions of tons of waste and debris buried in landfills that could potentially be repurposed and used as a resource. Companies could develop technologies and equipment to mine landfills and extract the valuable waste. Sorting equipment would need to be used onsite or at a nearby location to separate waste streams. Then it all comes down to logistics, networking, and business expertise to sell certain materials, or use them in your own manufacturing processes as a raw material. A lot of the plastic in landfills can exist for more than five hundred years, so after thirty years, it should still be intact and usable.

Metals can last in landfills for even longer and sustain great value. Aside from mining landfills and the economic opportunity in that, you could also create technologies or use existing tech to collect trash from street sides, waterways, the oceans, the Great Lakes, and population areas in general where people litter or pollute. Organizations like the Ocean Cleanup and 4 Ocean are innovating ways to collect trash from the oceans, shorelines, and beaches. They have already found ways to collect millions of tons of trash and raise money, and they are reusing and reinventing waste as a valuable resource. Their passion is fueled by the desire to save the planet from the environmental harm trash creates.

15. Start zero-waste, sustainable, or eco-friendly shopping centers, supply stores, grocery stores, restaurants, manufacturers, logistic firms, transportation firms, businesses, or any organization focused around a zero-waste platform.

There are not many box stores or physical shopping centers that have the sole focus of producing zero-waste or eco-friendly goods. However, there are online or digital shopping platforms like Net Zero Co., TerraCycle, Zero Waste Store, Zero Groceries, and Wild Minimalist that supply eco-friendly or sustainable consumer goods, with the emphasis of reducing any waste (especially plastic) and reusable packaging. The online stores might be able to also help limit carbon emissions from shipping, but the best solutions are usually to buy from your local store that sells only locally manufactured or agriculturally produced items. TripZero is a newer organization that helps you find more sustainable transportation and lodging methods and calculates the carbon footprint so you can offset it. There are midwestern local stores like Sawyer Home & Garden Center or Purple Porch Food Coop that focus on locally made goods and groceries and reducing waste through refillable containers. The problem is that stores like Sawyer and Purple Porch are few and far between. However, other storefronts are plentiful across the country. The following companies listed and total stores estimated are national and global brands with stores in most states and most average sized towns:

- 7-Eleven, at 71,000 estimated stores,
- McDonald's, at 38,000 estimated stores,
- Dollar General, at 16,000 estimated stores,
- Dollar Tree, at 15,000 estimated stores,
- Walmart, at 11,500 estimated stores,

- Aldi, at 11,000 estimated stores,
- CVS, at 10,000 estimated stores,
- Walgreens, at 9,000 estimated stores,
- AutoZone, at 6,000 estimated stores, and
- Whole Foods, at 500+ locations.

These stores provide myriad essential and convenient goods, but do they provide much in terms of zero waste or eco-friendly? Aldi and Whole Foods would be considered the most sustainable, with their emphasis on locally sourced, organic goods, and they sustainably form a holistic standpoint. Most of the others do not emphasize local, sustainable, or zero waste, and that is, in some ways, alarming when you consider the total number of store locations and amount of goods sold. The idea is that more towns of all sizes could have storefronts for zero-waste lifestyles, and some are popping up across the country already. The Zero Store has a storefront in the state of Washington, Empty Bin Zero Waste is in Ohio, and Zero Market is in Colorado. Recycle Away is an online storefront that provides categorized recycling, composting, and trash bins of all types, as well as sanitary supplies. Recycle Away would be a great physical storefront in any town. Here are some recommendations for any existing store or new store to implement to be more steward-like. These can be physical or digital storefronts.

- Achieve B certification through B Lab to improve stewardship,
- Adopt eco-shipping solutions and logistics for all shipped goods,
- Go recyclable-, reusable-, or returnable-packaging only,
- Implement converting to packaging free, plastic free, waste free,
- Implement net-zero waste practices, GHG emissions, and environmental offsets,
- Install renewable energy or convert your building to be more energy efficient,
- Open zero-waste lifestyle, supply and provision, or logistic stores,
- Provide and sell zero-waste consulting services (revaluing waste), or
- Sell sustainable, eco-friendly goods, and brands only.

16. Create a B-certified corporation, or start an entrepreneurial career focused on reinventing waste.

Now is the time to get into the business of reinventing waste. There are plenty of entrepreneurial, capitalistic, and consulting opportunities. Gaining the courage to start a business in general can be tough and is a huge accomplishment, but starting a business that is better for Earth is even better. Why not form a business that is for the greater good, better for the planet, and that still produces profits, like a B-certified corporation? You could also take it a step further and partner with 1 Percent for the Planet to donate money to causes helping save Earth. Life is full of opportunities to start your own business or join a company that is making a difference in the world. Now is the time, so join others in making a difference to help reinvent waste and reinvent Earth.

17. Build more Plastic Bank concepts or different waste banks, especially in populated or heavily trafficked areas.

Plastic Bank is an international organization making big impacts in reducing global plastic waste in the oceans, while also revolutionizing recycling ecosystems. Plastic Bank provides opportunities for individuals and organizations to offset their plastic footprints through credits. Also, Plastic Bank coined the term "social plastic," which is ethically recovered plastic material (which upholds to the United Nations' sustainable development goals) that transfers its value to communities in need. As a primary user of social plastic, one has a direct and traceable impact on helping to stop plastic ending up in oceans and improving lives. Plastic Bank is paving a new and innovative way that opens up significant opportunities. Why not create similar models for every modern waste streams or create a general waste bank concept that collects and uses all waste streams? These solutions could make tremendous positive impacts, especially in those population centers where waste and pollution tend to accumulate faster and in larger volumes. The concept of merging social equality with waste makes sense considering all the wasted value in the items being landfilled or causing pollution. Many people in the United States and across the world are looking for work, so why not give these people opportunities to help collect, sort, break down, and distribute material streams? Waste can and should become the new valued commodity.

18. Develop and invent new eco-friendly or more innovative plastics.

Eco-friendly or sustainable plastics can be evaluated and considered by:

- Renewable or regenerative resources/materials used
- Energy or material needed for manufacturing
- Fossil fuel dependency throughout the process or lifetime GHG emissions
- Biodegradable and compostable materials used
- How harmful a pollutant is if littered or landfilled
- Ease of shipping or general shipping footprint
- Reusability, recyclability, and reinvent-ability

As bulleted above, there are many variables to be considered when deciding which material, or which plastic, for that matter, has less of an overall environmental footprint. Having a longer lifetime and strong reuse and reinvent-ability value is sustainable. Balancing footprints and reinventing values seem to be great solutions for those deciding which plastic or material is eco-friendlier and more sustainable. There is an art in doing that, so continuing to rethink, reassess, and reevaluate will be essential. The Hemp Plastic Company is one company that is rethinking and reinventing the idea of plastic materials and manufacturing and navigating away from a petroleum dependency. There is a constant need for new materials, new plastics, and new products that can have a more-positive impact on humans and the environment. We need to continue to innovate and make materials, products, and life better. Great opportunities exist for those who want to develop and invent more eco-friendly and innovative materials, products, or plastics.

19. Procure sustainably farmed, sourced, and eco-friendly raw materials or feedstock in the manufacturing of products.

Modern sustainable-focused businesses in either manufacturing or producing some type of product need to deeply consider procuring only sustainably farmed, sourced, and/or eco-friendly raw materials (feedstock). What does that mean? *Sustainably sourced* or *eco-friendly* means a raw material that is procured or farmed from a location that bio-regenerates quickly. Also, the material should be more abundant and easier to collect footprint-wise and be procured by practicing environmentally safe, controlled, and ethical collection or removal. Some innovative and proactive companies that are doing just this are EarthHero, Who Gives a Crap, Pela, and Green Toys. Model after these companies and join the movement of innovation.

20. Develop more plant-based (bio) materials, compostable packaging, and bio solutions.

The evolution of technologies and continuous innovation have paved ways for new compostable, plant, or bio-based packaging and products. Companies like World Centric, Seventh Generation, Earth Choice, ECO Products, Yes Straws, and Native are doing just that. The natural world provides many materials that are strong, durable, long-lasting, and that regenerate quickly. Humans have engineered and scientifically merged materials to invent new materials not naturally found on Earth. The problem is that if new materials like plastic aren't handled correctly, they can be environmentally disruptive. It would make sense to keep materials in more of a natural state, so that they can easily degrade back into nature and contribute to an ecosystem of regenerative materials. Companies developing plant-based materials, compostable packaging, and various other bio solutions are changing the way humans think about materials, which opens the door to new possibilities. Explore the world and learn about natural materials that already exist and use those material as manufacturing feedstock.

21. Invent waste-to-fuel, -energy, or new-material innovation.

There are companies today that are paving the way to innovate waste in different ways. One company is Brightmark, which is turning plastic into fuel or new, usable materials for manufacturing feedstock, and are creating renewable natural gas from food, dairy animals, and organic waste digesters. Brightmark is strategically planning to build plants across the country that will collect a large portion of the plastic waste that is quickly accumulating. The model that Brightmark is formulating is creative and innovative and provides inspiration for others to embark on a similar journey.

22. Start waste or litter cleanup (recycling) and maintenance companies. Examples may include hiring a firm to clean up sections of highways, waterways, cities, or beaches.

Most organizations with big buildings typically have a janitor or maintenance person on staff, or they hire a firm to take care of those essential duties to keep a building clean. Why not apply this same concept to Earth in general, especially those densely populated areas? One could start a company focused on waste removal and cleanup from any place on Earth and get paid to do it. On top of getting paid to do it, one could potentially take that waste and recycle it somewhere for additional revenue. Not only does this provide more job opportunities in stewardship, but it can help clean up the planet and contribute to an improved sustainable waste management infrastructure.

23. Build eco-parks or sustainability parks, sustainability think tanks, or incubator hubs, community centers and recreational areas.

Imagine a place where sustainable business practices, business incubation and innovation hubs, recycling centers, renewable energy plants, nature or conservation centers, recreation, and community gathering all come together as one. That is what an innovative eco-park or sustainability park could be, and that is exactly what Growing Green is developing in Indiana. The idea is a real innovation breakthrough, reinventing the way we can think about modern parks. Parks do not need to be only about nature and recreation; they can include sustainable solutions to make the world a better place and create economic impact opportunities. These parks could also provide more equity and inclusion to its workers and community.

24. Create a composting company to build more innovative products for composting.

There are modern methods to composting at your home or business, on a small or big scale, and indoors or outdoors. Worms, probiotics, bacteria, insects, and bugs can all break down organic waste. Put your organic waste (anything that grows from soil) into some type of receptacle or contained area, mix in some living organisms, and out comes compost soil that is great for gardening and agriculture in general. Companies are already making innovative products or systems for composting, but there is plenty of potential for new ideas or concepts. The increase of compostable products in the world should allow for additional materials to be composted along with all our organic waste. From an infrastructure standpoint, local governments, waste haulers, and landfill companies can partner to discover new methods for organic waste recovery and separate it and create state-of-the-art composting systems. The Solid Waste Association of North America could help educate all stakeholders and encourage the separation of organic waste to increase composting.

25. Construct long-lasting quality products that are made with eco-friendly materials. Make free or low-cost repairs and returns a pivotal part of the business platform.

Build or produce items of great quality that endure longer results in more sustainability. The amount of energy and types of materials used matters greatly, but if something is made to last longer, it reduces future energy or material needs. To make free or low-cost repairs and returns a part of the business model is smart as well. If an item can be repaired using a minimal amount of energy and minimal new material, that is true stewardship and more sustainable than having to buy a brand-new item that used more material and energy to make. Companies like Patagonia's Worn Wear concept, Hestra, and Genius Phone Repair are great models for making both quality and repair the essence of the business platform. If you're thinking about starting a new business of producing products, deeply consider making only quality and long-lasting products, as well as repair services, the core of your business. It should lead to greater customer loyalty and a large following.

Step 4—Restore

Restore Ancient Natural Systems and Pave a Sustainable Roadmap for Life

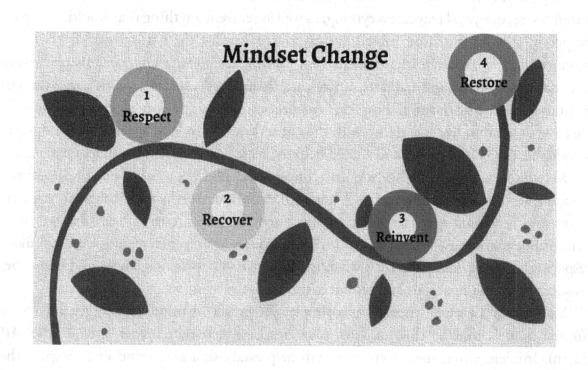

Closing the Loop and Restoring Earth

The fourth and final step is to restore, which completes the cycle (respect, recover, reinvent, and restore) for creating a sustainable system or framework for living in balance with our Earth. Can we create a way of life that is more in balance with Earth's natural systems? That may be the biggest question of the current era, and it will take significant effort by all humans to restore the planet. The human spirit has brought about amazing technologies, for better or for worse, and now is the time to fully use research, knowledge, innovative technologies, and science as never before to reverse the negative impacts of the current waste crisis.

Step 4, restore, really brings to light the idea that if humans can respect natural resources, recover, and recycle everything, as well as reinvent anything that would otherwise be potentially landfilled or littered, we can indeed restore the natural world. Nature is more resilient than anything and can heal itself, if it's given a chance. History shows that natural systems and habitats can restore themselves, especially with some human assistance. Scientific research shows that resources are finite and only regenerate so fast, so we have to learn to live in balance with the natural world. To look at resources and waste as valuable material is smart and should help to create better systems to manage material.

Materials management is a modern concept and philosophy that has been gaining traction around the world and will hopefully continue to do so with great velocity. Mastering materials management can help Earth heal and restore itself. But we must better understand how long it takes for those resources to regenerate and use them more responsibly. To be good stewards, we must master materials management and consume, reuse, and recycle materials and resources more wisely.

Sometimes all that is needed is a simple, yet powerful mindset change. That new mindset shift could be this four-step plan: respect, recover, reinvent, and restore (4R Earth). Implementing these four steps will help establish a new mindset to respect the natural world and recover waste and resources to the best of our abilities. We should reinvent waste to use it as a valuable resource and to restore the Earth so that humans can live more in balance with the planet and assure that it provides for all future generations. Completing the final step, restore, would close the loop and allow things to come full circle. By restoring Earth's ancient natural systems (endangered forests and more), we create a sustainable roadmap for humanity to live by, living in balance with Earth now and forever.

1. **Respect** Earth and understand its natural systems
2. **Recover** resources (waste) by improving industry infrastructure
3. **Reinvent** waste as the most undervalued resource/material on Earth
4. **Restore** ancient natural systems and pave a sustainable roadmap for life

A Mindset Evolution or Revolution

One significant way to reverse negative impacts of poor waste management and to truly reinvent waste globally, is through a positive mindset change or shift among all of humanity: ourselves, our families, cities, businesses and organizations, schools, health care, and governments. Learning to live in balance with Earth and its natural systems and changing traditional mindsets would help individuals build more sustainable lifestyles. The more people learn to live sustainably, the more it should inspire others that they can do the same, and that momentum could really improve sustainable waste management across the country and the planet.

One way to think about this idea of mindset change is to envision a plant, like the one below. A tree or plant grows almost magically from nutrient-rich soil from the earth or ground. Then, it takes what it needs to survive on a daily basis. After that, it is creative and strategic, protects itself and sometimes others, is innovative and resilient, lives as long as it can, and finally, slowly dies (decomposes), ultimately giving its nutrients back to the soil. It's a natural cycle.

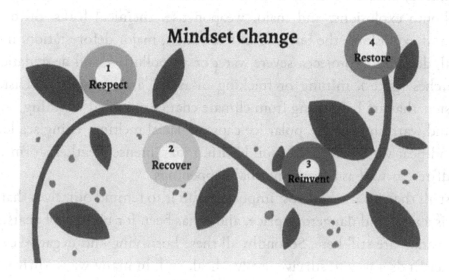

Those leftover nutrients from dead plants and trees go back to the soil to give new plants and trees an opportunity to thrive, and that is critical to the circle of life for most trees and plants. Humans have the opportunity to copy or to be inspired by nature, with some intensive effort, to have little to no impact on the planet so that future generations can also have a nutrient-rich life. *Biomimicry* is a term used by the Biomimicry Institute. The idea is to be inspired by nature to solve business or organizational issues and to innovate or evolve.

Another way to think about this mindset shift is to think of giving nature time to heal and restore itself and to be inspired and help preserve the planet so it can last forever.

It makes sense to respect the planet, recover materials or resources, and reinvent the way we think and live to help give the planet time for restoration. Is humanity willing to join together in a true mindset change, movement, evolution, or revolution? Only time will tell how things evolve from here. If proper steps are taken now, humanity could regain a sense of balance with Earth and be able to help it recover a sustainable baseline. We have only one livable Earth that we know of, or have access to, so we should try our best to live in proper balance with it. Earth is resilient, and if respected, it could have time to restore itself so that it provides life for all future generations.

Our Resilient Planet

Natural disasters, such as meteorites, ice ages, hurricanes, cyclones, tsunamis, floods, earthquakes, landslides, erosion, volcanoes, dust storms, tornadoes, and wildfires have wreaked havoc on the planet since the beginning of time, and most still occur on a daily basis somewhere on the planet. On top of these natural disasters, there are anthropogenic (human-induced) disasters. Some of these disasters are ozone hole creation or ozone depletion, bomb explosions, and major weapon use, chemical leaks, oil or gas (fossil fuel) leaks and explosions, the burning of fossil fuels, major deforestation, mountain or soil removal, demolition projects, severe water or air pollution, soil degradation, oceanic garbage patches (gyres), mining or fracking disasters, and great fires. Last, there are newer disasters that are happening from climate change or global warming, such as coral bleaching and warming oceans, polar ice caps or glacial melting, rising sea levels, severe droughts and heat waves, declining soil health, more intense weather storms, and more serious wildfires, as well as other abnormal happenings.

Why list all these disasters? Most importantly, it is to remind ourselves that the planet we inhabit is a wild and dangerous place, and it has been for billions of years, yet we (all life on the planet) are still here. Secondly, all these horrifying and negative events occur, yet the planet finds a way to survive, evolve, heal, and, in many ways, thrive. The Earth is actually very resilient and is able to regenerate. Therefore, we can remain hopeful that even though we may be negatively impacting the planet now, we can change course and give Earth enough time to heal and allow itself to restore to a more sustainable balance. A sustainable balance means humans consume only what is needed for a good life, without harmfully impacting the environment and without comprising the ability of future generations to also live a good life.

As mentioned in step 1, respect, data shows that humans are living out of balance with Earth's natural systems, and if our actions are not reversed, the planet could be in grave danger. We are wasting money and harming the planet by dumping more and more materials into our landfills every day, as well as continually polluting the land and sea with trash. Climate change is also worsening daily, and greenhouse gas levels like carbon are reaching all-time highs, which now is likely accelerating the warming of our planet. Many scientists and researchers are proposing that the geological epoch we live in now, the Holocene, also known as the Anthropocene—a word that means human activity—has been the dominant influence on climate and the environment. Some believe we are currently dealing with the sixth mass extinction event, where flora and fauna loss or extinction is reaching record highs. The question we should be asking ourselves is, why are we venturing down the current pathway that is knowingly putting our planet at risk?

Holocene or Anthropocene?

One easy solution is to follow the four-stepped plan: to respect the natural world, recover resources to the best of our abilities, reinvent waste to use it as a valuable resource, and to restore the Earth so that humans can live more in balance with it, assuring that it provides for all future generations. Not only will this help people realize the true value of waste; it can help restore Earth's ancient natural systems and pave a sustainable roadmap for individuals to live a more healthful life. Once waste management is fixed from a systematic and sustainable standpoint, we can be optimistic that it could also help reduce the negative impacts of climate change. Living more in balance with the Earth includes minimal negative environmental impacts, thus a potential slowdown of climate change. So, isn't reinventing waste seemingly a safer path?

Stop Emissions from Traditional Waste Management

To understand more about the depth of negative environmental impacts of waste, we can look at metrics like estimated GHG emissions. See Figure 46 below, which shows emissions from 2011 through 2017. It is important to note that the numbers do not include biogenic emissions, transportation, or logistical emissions, emissions from decaying trash around our highways or waterways. The clear finding from Figure 46 is that negative environmental emissions continue to be moderately large year over year, in millions of metric tons. Clearly, there is a tremendous need for change and improvement.

Figure 46—Reported Emissions by Waste Subsector (2011–2017)

Waste Sector	Emissions (MMT CO$_2$e)[a, b]						
	2011	2012	2013	2014	2015	2016	2017
Total Waste Sector	**114.9**	**115.0**	**111.2**	**111.8**	**110.3**	**107.6**	**105.6**
MSW landfills	93.8	94.4	91.1	90.8	89.7	86.9	86.5
Industrial wastewater treatment	2.6	2.1	2.2	2.6	2.1	1.9	1.9
Industrial waste landfills	8.9	8.7	8.0	8.5	8.5	8.6	8.1
Solid waste combustion	9.6	9.8	10.0	9.9	10.1	10.2	9.2

[a] Biogenic emissions of CO$_2$ are not included in the CO$_2$e emissions in this table. As landfill gas recovered from MSW landfills and industrial waste landfills is considered biogenic, CO$_2$ emissions from the combustion of landfill gas are not included in the CO$_2$e emissions in this table. Biogenic CO$_2$ emissions from the combustion of the biogenic fraction of MSW are also not included in the CO$_2$e emissions in this table.

[b] Totals may not sum due to independent rounding.

Source: EPA, Greenhouse Gas Reporting Program, Waste Sector, 2019 Report.

On top of the ongoing GHG emissions, humans are still consuming biodiversity faster than it can reproduce, as we reviewed earlier, all while people and organizations continue to bury trash that has no demand yet (willing consumers or buyers). Not only are these three key issues an enormous economic miss, the planet (and humans) will likely not survive much longer on the current trajectory or pathway. This begs the question, why aren't people acting more fervently to avoid a potential major catastrophe?

For those who want to make a real difference, a positive impact, and help the planet reinvent itself so that it can sustain forever, the time to act is now. One major need and a quick win is to educate one another and empower others to give Earth time to heal and regenerate. Second, we should take only what we need to survive and thrive, while giving nutrients back to mother nature so that those same resources consumed can then regenerate. That regeneration potential provides future generations the same opportunity to survive and thrive. This is a similar path to what E. F. Schumacher discusses and proposes for all earthlings (consumers) in his book *Small Is Beautiful*. The big idea is that consuming less and living more frugally provides greater health and happiness to all the life on Earth, particularly flora and fauna. Another good idea to bring this about is minimalism. One example of a minimalist is someone who chooses to live in a smaller-than-average home, resulting in less consumption of energy, resources, and goods. Another example is buying only one car for your entire family, thus encouraging carpooling, more walking, taking a bike, or using public transportation for travel and commuting. The idea of minimalism can be adopted and applied in any household, business, organization, or government office, and can lead to less waste and more opportunity to reinvent waste.

Rewilding the Planet Concept

In the early 1990s, Dave Foreman coined the term *rewilding*. Dave cofounded Earth First! in 1980, and the Wildlands Project in the 1990s, which eventually evolved and became the Rewilding Institute in 2003, which Dave founded. Throughout his career, Dave has worked with large environmental organizations, such as the Wilderness Society and the Nature Conservancy. He has authored books including *Rewilding North America*, *Take Back Conservation*, and *The Big Outside*. Dave is known as one of the most charismatic and influential leaders in the fields of conservation and wilderness restoration. To Dave, the term *rewilding* is wilderness and apex carnivore restoration. The movement has inspired biodiversity protection and restoration around the world. Principles like conservation biology, wildlands network design, and continental-scale conservation are some of the very important frameworks that Dave and his team follow (The Rewilding Institute n.d.).

REWILD Earth

So, how does the concept of rewilding the planet align or have any connection to reinventing waste? Rewilding the planet is a concept concerning taking care of the planet and being a good steward for nature and all things living. Reinventing waste is about stopping the waste crisis that we have been in, and instead, better using waste as a resource for living in balance with the planet and its available biocapacities. So, you could say that the need to reinvent waste and to rewild the planet in parallel, as they are both very important.

As discussed in step 1, respect, the last 250-plus years marks a period of great significance. Three major human sequential revolutions have accelerated human development faster than ever before, and now the health of the planet is grim and Earth can't sustain itself. A majority of the world lives as though there are 1.7 Earths in terms of available biocapacity, and here in the United States, it's even worse. We are living in what is known as the Anthropocene period, which many are calling a climate crisis and the sixth mass extinction of flora and fauna. But as we discussed, Earth is resilient and has survived so much turmoil throughout history. So, we must have faith that anything is possible. The opportunity is there, and now it's up to us to restore ancient natural systems and reestablish a balance with the planet and live harmoniously with the natural world.

To rewild the planet, we must act courageously to preserve ancient forests and rainforests, oceans and coral reefs, wetlands, prairies, and meadows, and other ancient biodiverse places on Earth. We must be respectful and responsible in our removal of Earth's minerals and resources, as biodiversity is absolutely essential. In terms of agriculture, being efficient and resourceful in terms of land use and total consumption will be critical. Reducing our human footprint on air, water, and soil to a new all-time low will potentially be the greatest challenge in human history, but a worthy one we must act on. As good stewards, we have the ability to learn about botany and how to reforest and replant to help encourage new growth in those ancient zones that are important to the natural flow of the planet. Rewilding conceptually is also about being inspired to innovate and continuously preserve our natural world. To reinvent waste, one can learn from the rewilding mindset.

Invite Innovation and Evolution to Support Restoration

Technology is seemingly always changing and evolving, as the human spirit grows in wisdom. Wisdom and possibilities are passed on through generations and provide opportunities for future generations to continue to evolve and become more intelligent throughout time, and to reach new heights and achieve new milestones. This process can be called innovation. Innovation has propelled human civilization to amazing feats in terms of technological development. So, one idea is to use that tech innovation and evolution to reinvent waste and create hope for a sustainable future. A sustainable future is one that gives the Earth time to heal and restore biodiversity, allowing humans to coexist with nature, having a minimal impact on the planet.

In the world of recycling, technology has already changed the game in terms of logistics, sorting, separating waste and redistributing waste (material) streams to desired parties. However, the opportunity to further use the digital and technological revolution seems immense. In terms of technological and digital evolution, there are major opportunity areas for the reinventing-waste industry to tap into. The following are all excellent innovation verticals happening simultaneously:

- Robotics, automation, and programming
- Sophisticated machine evolution
- Machine learning and artificial intelligence
- New-tech advancement, exploration, and creation
- High-speed internet networking and connectivity
- Data-storage and capability and server capacities
- Software and hardware improvement
- Application, smart tablet, or cell phone development

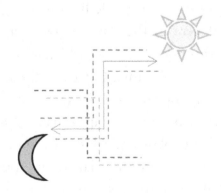

- Clean tech, renewable energy, and energy storage
- Electrification of transportation and alternative transportation
- Smart logistics, shipping, and warehousing

Each innovation vertical provides a different set of opportunities for individuals, companies, and organizations to incorporate into reinventing-waste mindsets. The human brain is truly amazing and continues to move boundlessly through preexisting barriers of what was once thought to be impossible. In today's world, there is so much to be thankful for, as technology can be used to create good in the world for future generations. The key is to create good while absolutely minimizing human impact on nature and the Earth as a whole. Inviting innovation and the evolution of technology can be a force for good. It is up to the inventors and innovators to do it the right way, reinventing waste rather than creating waste that has no use.

New Sustainable Roadmap for Life on Earth

It will take a strong collective effort and vision from all of humanity to restore ancient natural systems and pave a more sustainable roadmap for life. A great plan has been laid out by the United Nations from the 2015 General Assembly. The gathering created an in-depth plan with key focus areas for the world to consider as imperative for survival of the planet and all things living. A total of seventeen key goals were created (Figure 47), along with subgoals and strategies related to each vertical.

The impacts—both positive and negative—from waste and the opportunities to reinvent waste can truly be tied to each one of the seventeen sustainable development goals (SDGs). They range from equality, well-being, environmental, and education, to innovation and general human development. Goal twelve, responsible consumption and production, might be the one that stands out the most in terms of reinventing waste. The main intent of SDG twelve is to ensure sustainable consumption and production patterns of Earth's natural resources and biodiversity.

Figure 47—The Seventeen Sustainable Development Goals

Source: United Nations Sustainable Development Goals, 2015 General Assembly.
Link: https://www.un.org/sustainabledevelopment/. The content of this publication has not been approved by the United Nations and does not reflect the views of the United Nations or its officials or Member States.

Seventeen Goals for People and for the Planet:

Heads of State and Government, senior UN officials and representatives of civil society gather[ed] in September 2015, as part of the 70th session of the UN General Assembly and have adopted the Sustainable Development Goals (SDGs). These objectives form a program of sustainable, universal and ambitious development, a program of the people, by the people and for the people, conceived with the active participation of UNESCO, or the United Nations Educational, Scientific and Cultural Organization.

The Sustainable Development Goals are a universal call to action to end poverty, protect the planet and improve the lives and prospects of everyone, everywhere. The 17 Goals were adopted by all UN Member States in 2015, as part of the 2030 Agenda for Sustainable Development which set out a fifteen-year plan to achieve the Goals. Today, progress is being made in many places, but, overall, action to meet the Goals is not yet advancing at the speed or scale required. 2020 needs to usher in a decade of ambitious action to deliver the Goals by 2030.

With just under ten years left to achieve the Sustainable Development Goals, world leaders at the SDG Summit in September 2019 called for a Decade of Action and delivery for sustainable development, and pledged to mobilize financing, enhance national implementation and strengthen institutions to achieve the Goals by the target date of 2030, leaving no one behind.

The UN Secretary-General called on all sectors of society to mobilize for a decade of action on three levels: **global action** to secure greater leadership, more resources and smarter solutions for the Sustainable Development Goals; **local action** embedding the needed transitions in the policies, budgets, institutions and regulatory frameworks of governments, cities and local authorities; and **people action**, including by youth, civil society, the media, the private sector, unions, academia and other stakeholders, to generate an unstoppable movement pushing for the required transformations. (United Nations n.d.)

Source: United Nations.

The SDGs encompass everything imaginable in terms of holistic sustainability. They lay out the roadmap and encourage the world to live in balance with nature and allow for humans to still try and thrive on this planet. The biggest hurdle is, and will be, getting enough people on board to use and follow the roadmap, and to stick with it. If most of the planet (people) can align and act swiftly, the new wave of positive impacts should offset the harm and damages already done. Perseverance will be critical, and the patience of many will be severely tested. Experts are ringing the bells in terms of raising awareness and creating a sense of urgency for good stewards to act, and to act now. Now is the time. Will you stand up for what is right and join the movement? Use all seventeen of the SDGs, especially number twelve, to reinvent waste and be a good steward for Earth.

Tips for Stewards: Seven Ways to Restore Earth

1. Build a "Nothing is Impossible" Moxie
   - ☐ Have you watched *14 Peaks* on Netflix yet? It is possibly the most inspiring and motivational film ever released. Long story short, a Nepalese climbing expedition called "Project Possible" broke numerous world records, one of which was climbing all fourteen peaks above eight thousand meters high (the highest mountains in the world) in just over six months. They broke the previous record by a long shot, just under eight years. Nirmal "Nims" Purja was the expedition leader and creator. Many experts and close friends told him his project was impossible and insane, but Nims assembled a team of like-minded fearless individuals, and together, they accomplished what was thought to be impossible, climbing the world's highest fourteen peaks in six months and six days. According to Nims, the driving motivation for creating and conquering "Project Possible" was to inspire the world and the next generations that literally anything is possible, as well as encouraging a stronger fight against climate change. Thank you, Nirmal Purja, for proving that nothing is impossible, and that anyone can attain that moxie.

2. Materials Management Mindset Development
   - ☐ Embracing and developing a materials-management mindset could be the mighty mindset change that is so needed for the United States and for the planet. If every one of us handles materials (natural resources, biodiversity, manufactured or produced goods, and so on) with deep care and consideration, likely most waste could be diverted. The old saying, "One person's trash is another one's treasure" still rings true because any material stream in large enough quantities is considerably valuable to someone or some organization. It may take time, resources, strategic systems, and processes (sorting and separating) to collect large enough volumes of particular material streams for resale or redistribution; however, the economic or environmental benefits are typically worth it. TerraCycle proved that this mindset is valid with its cigarette butt recycling program.

3. Continuously, or Incrementally, Learn, Innovate, Improve, and Evolve
   - ☐ It might sound simple, but continuously learning, innovating, and improving can be a grind. Humans are wired to take the easy route or most convenient path. But the easy or convenient path isn't always best for the planet. We need to encourage each other to continuously learn, innovate, improve, and evolve in balance with the planet. If we let our individual creativity flow, we can tap

into that thirst to continuously improve on what the generations before us innovated in terms of sustainable human development. New-Tech, or methods like upcycling, might get us to that sustainable balance, but we must consider incrementally sticking with goals to continue that impactful momentum. The Incremental Development Alliance is one organization leading a movement and mindset concept for good stewards to embrace incrementalism for more sustainable development.

4. Embrace Circularity and a Circular Economy
   ☐ Embracing circularity methods like the circular economy can help good stewards (consumers or producers) live more harmoniously with the planet. Step 4, restore, is about closing the loop and thinking about sustainable systems. Another way of looking at it is thinking with products' end of life in mind. However, with closed-loop systems of production, distribution, and consumption, materials continuously flow systematically with no or very little waste. Sustainable materials and systems embrace circularity and close the loop. Designing and manufacturing with such concepts in mind as EPR and DfE embrace circularity very well and can be embedded into the fabric of any organization. If humanity is going to live out a true circular economy, we must embrace circular methods.

5. Live in the Spirit of a Native American
   ☐ The mind, body, and spirit of Native Americans is very well-connected to Mother Earth. As we discussed earlier, most Native American cultures live with a mindset of how their decisions or lifestyles impact future generations. Native American cultures, especially the ancient ones, teach us that zero-impact lifestyles are feasible. Before the leave-no-trace concept was popular in hiking and camping, Native Americans literally left no trace or had minimal impact on the planet for thousands of years. Living in the spirit of a Native American might be enough to rewild and restore the planet so that it can be sustainable for future generations, allowing them to survive and thrive.

6. Act Swiftly on the Sustainable Development Goals
   ☐ A decade of action is critical, especially according to the United Nations. The United Nations has laid out seventeen sustainable development goals for us to follow and focus on for people (equality for all), the planet (environmental protection) and profit (economic opportunities). Global experts and the UN leaders believe this decade is more important than ever in terms of major environmental tipping points on the natural world that will have a negative

impact on people and profits, if not reversed. We must act swiftly and collectively. Let's use goals like the seventeen SDGs.

7. Create Your Own Pledge to Reinvent Waste
    ☐ Momentum can grow when people start pledging, signing petitions, and collectively trying to achieve the same goal. You could create your own or sign a pledge to achieve any goal or goals around reinventing waste. Pledging shows you're serious, responsible, committed, and willing to be accountable for achieving or not achieving a set goals. Achieving goals boosts moral, confidence, and even inspires individuals to set newer and more aggressive goals. One recent popular pledge is the "Climate Neutral Now" Pledge by the United Nations, which raises awareness and is creating positive momentum for reducing global carbon and GHG emissions.

A Destination Unknown

No one knows what the future holds for the planet, but the current trajectory does not bode well, especially in terms of climate change. It could be argued that we each have a duty and responsibility to be good stewards for the planet and to protect it for future generations. Greta Thunberg is a popular climate activist who is preaching to the leaders of the world just that—to have the courage to fight climate change and alter the path we are on. The COP26 conference recently took place, which is a pivotal movement in the fight against climate change.

> In November 2021, the United Kingdom, together with partners in Italy, hosted an event many believed to be the world's best last chance to get runaway climate change under control. COP26 was the 2021 United Nations climate change conference. For nearly three decades the UN has been bringing together almost every country on earth for global climate summits—called COPs—which stands for 'Conference of the Parties.' In that time climate change has gone from being a fringe issue to a global priority.
>
> This year was the 26th annual summit—giving it the name COP26. With the United Kingdom as President, COP26 took place in Glasgow, Scotland. In the [run-up] to COP26 the United Kingdom worked with every nation to reach agreement on how to tackle climate change. World leaders arrived in Scotland, alongside tens of thousands of negotiators, government representatives, businesses, and citizens for twelve days of talks.

Not only was it a huge task, but it was also not just yet another international summit. Most experts believe that COP26 had a unique urgency.

COP21 took place in Paris in 2015. For the first time ever, something momentous happened: every country agreed to work together to limit global warming to well below 2 degrees and aim for 1.5 degrees, to adapt to the impacts of a changing climate and to make money available to deliver on these aims. The Paris Agreement was born. The commitment to aim for 1.5 degrees is important because every fraction of a degree of warming will result in the loss of many more lives lost and livelihoods damaged. Under the Paris Agreement, countries committed to bring forward national plans setting out how much they would reduce their emissions—known as Nationally Determined Contributions, or 'NDCs.'

They agreed that every five years they would come back with an updated plan that would reflect their highest possible ambition at that time. Glasgow was the moment for countries to update their plans. The [run-up] to the summit in Glasgow was the moment (delayed by a year due to the pandemic) when countries updated their plans for reducing emissions. But that's not all. The commitments laid out in Paris did not come close to limiting global warming to 1.5 degrees, and the window for achieving this is closing. The decade out to 2030 will be crucial. So as momentous as Paris was, countries must go much further than they did even at that historic summit in order to keep the hope of holding temperature rises to 1.5 alive. COP26 had to be decisive.

COY16 was the sixteenth United Nations Climate Change Conference of Youth.

Organised in collaboration with YOUNGO, The Official Youth Constituency of the United Nations Framework Convention on Climate Change (UNFCCC), it was one of the largest entirely youth-led global youth climate conferences in the world. The conference took place from October 28–31, 2021, days before the annual United Nations Climate Change Conference, also known as Conference of the Parties, in the same host country as the COP.

COY served as a space for capacity building and policy training, in order to prepare young people for their participation at COP and their life as local and international Climate advocates. The Conference was organized by seven Working Groups and Coordinators in 152 Countries whose main objective was to gather local Youth Statements and raise the voices of the global youth to the United Nations decision making process.

COY16 culminated in the Global Youth Position Statement, representing the views of over forty thousand young people worldwide. The statement was profiled at COP26 on Youth & Public Empowerment Day. (COP26 2021)

The reason for discussing COP26 and the issues around Climate Change is to raise a sense of urgency about taking care of the planet. No matter your religious and political beliefs or your ethnicity, you do have one thing in common with every other person on the planet: you are an earthling, and you can be a good steward for Earth! No matter what your profession is, you are still an earthling. We all inhabit the same planet, and we are all trying to survive and thrive on it. We must continually learn from each other and encourage each other to be better stewards every day.

The science, data, and the statistics that are being reported daily and annually showcase the gravity of our situation on planet Earth. Climate change is real, and it involves waste. Although waste may not be the top culprit for the planet changing so drastically, it is a very complicated issue and certainly has an enormous negative impact on the planet. As population grows, the consumption of goods and resources will only increase; therefore, we must implement solutions now. It will take a collective effort from each state, country, and world to make the necessary difference. Together, if we mitigate waste to the best of our abilities and revalue it, we likely can make a very positive impact on the planet and help reduce climate change.

Fortunately, there are solutions, movements, and initiatives you can take part in right now. Or you can innovate, create, or invent your own ideas like the four-step plan 4R Earth. The four-step plan will help add to the foundation of modern sustainability and give birth to fresh mindsets that will constructively reinvent waste and make it a valuable resource or feedstock for production and consumption. Waste is likely the most undervalued resource of our time, and now is the time to reinvent it and value it like responsible stewards. As stewards, we should try to regain balance with Earth and restore it. We must try our best to live in proper balance with Earth and allow it to heal and regenerate its biocapacities. The Earth is resilient, so if we collectively respect it, it should be able to restore itself so that it can provide life for all future generations. This study aims to raise awareness, provide ideas, and help create new mindsets to live by for individuals who want to be better stewards and make the planet a better place. This is a steward's guide for reinventing waste.

Hope Is the Way

It is easy to feel doom and gloom about topics like waste, overpopulation, available biocapacity being depleted, landfills getting bigger, plastic, or general pollution becoming increasingly worse in oceans, the polar ice caps melting, deforestation, or destruction of ancient natural areas of the planet that store large amounts of carbon, fires, and drought wreaking havoc across the globe, mass extinction, and other climate change outcomes. So, how do we have hope? Fortunately, we are wired to survive and pursue ultimate happiness, and for that reason, we remain hopeful and try our best most days. So, as Yoda of the Star Wars movies says, "Faith, you must have."

New initiatives, movements, and policies are being implemented daily across the country and the planet, and that gives me hope. What also gives me hope is that every earthling on the planet can let his or her own creativity or art flow and make his or her own positive impact on the planet. One positive impact could be taking a waste stream and using it or reinventing it as a feedstock for new innovation. People can live a new type of zero-impact or net-zero lifestyle, where sustainability is maximized and ecological footprint is significantly minimized. You can sign the "Climate Neutral Now" pledge, from the United Nations Framework Convention on Climate Change to commit to going net zero on greenhouse gas emissions by or before 2050. You can take the pledge as a person, family, or organization. All you have to do is measure your GHG emissions annually, plan, and implement actions to reduce them entirely. Finally, you have the option to become a sustainability champion and to help other climate enthusiasts reduce global emissions.

Each day is an opportunity to learn more about the planet and how humans can live more harmoniously with it. We can wake up each day and choose to act as better stewards for our Earth. You can pursue a journey of continuous learning, teaching, and implementing ideas to make the world a better place. You can be a good steward for our Earth and can have fun with it, and even make a business out of that idea. Just make sure you enjoy it because our time on this planet is short in the grand scheme of time and the existence of our Universe.

Thanks for reading, I hope you learned something and were able to get some new, inspirational, and thought-provoking ideas for reinventing waste. Please connect with me if you would like to discuss a potential solution or have a business idea. I would love to hear about it!

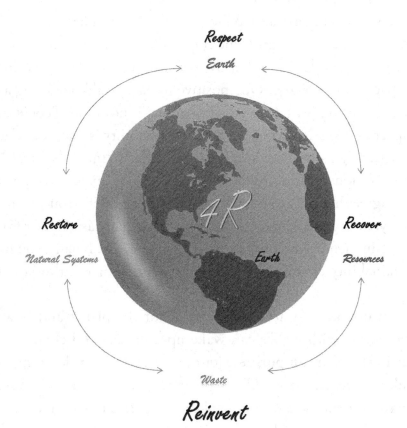

Respect

Earth

Restore

Natural Systems

Recover

Resources

Waste

Reinvent

References

ACLCA. n.d. "About ACLCA." Last Accessed 2023. https://aclca.org/about/

Alliance for the Great Lakes. 2018. "Great Lakes Plastic Pollution." Last Accessed 2023. https://greatlakes.org/great-lakes-plastic-pollution-fighting-for-plastic-free-water/

American Biogas Council. 2020. "What is Anaerobic Digestion?" Last Accessed February 21, 2018. https://americanbiogascouncil.org/resources/what-is-anaerobic-digestion/

American Disposal Services. n.d. "A Brief History of Recycling." Last Accessed November 19, 2019. https://nerc.org/news-and-updates/blog/nerc-blog/2019/11/19/a-brief-history-of-recycling

B Lab. 2018. "About B Corp Certification." Last Accessed 2023. https://www.bcorporation.net/en-us/certification

Bass, Dar. 2018. "Recycling Thoughts and Innovation for Kent County Michigan and Beyond." Interviewed via Phone by Tyler Kanczuzewski

BBC Science Focus. 2022. "The Life Cycle of a Star: How Will Our Solar System End?" Published October 21, 2022. https://www.sciencefocus.com/space/the-life-cycle-of-a-star-how-will-our-solar-system-end/

Bradbury, Matt. 2017. "A Brief Timeline of The History Of Recycling." Published May 20, 2017. https://www.buschsystems.com/resource-center/page/a-brief-timeline-of-the-history-of-recycling

Bradley, Luanne. 2009. "A Handy Reference Guide to the 20 Greenest Materials." Ecosalon. Last Accessed 2023. http://ecosalon.com/a-handy-reference-guide-to-the-20-greenest-materials/

Center for Health, Environment & Justice. 2015. "Landfill Failures, the Buried Truth." Published June 2015. https://chej.org/wp-content/uploads/Landfill-Failures-PUB-0091.pdf

Christopher, Norman. 2012. *Sustainability Demystified!: A Practical Guide for Business Leaders and Managers.* Grand Rapids: Principia Media.

Clark, Teresa. 2017. ""People, Planet, Profit" – It's Not A Triple Bottom Line." Greenpath Recovery, Blog. Published March 10, 2017. Last Accessed 2023. https://greenpathrecovery.com/blog/people-planet-profit-not-triple-bottom-line/

Congressional Research Service. 2017. "Federal Land Ownership: Overview and Data." Congressional Research Report, Prepeared for Members and Committees of Congress (March): 2. https://crsreports.congress.gov/product/pdf/R/R42346/15

COP26. 2021. "What is a COP?" UN Climate Change Change Conference Glasgow 2021. Last Modified November 2021. https://ukcop26.org/uk-presidency/what-is-a-cop/

Cross, John-Michael. 2013. "Fact Sheet|Landfill Methane." Environmental and Energy Study Institute. Published April 26, 2013. https://www.eesi.org/papers/view/fact-sheet-landfill-methane

Dawes, Jordan. 2018. "Kent Trash Reuse Project Worth $500M." *Grand Rapids Business Journal,* August 10, 2018. https://grbj.com/news/kent-trash-reuse-project-worth-500m/

Earth Day Network. 2018. "The History of Earth Day." About Us. Last Accessed 2023. https://www.earthday.org/about/the-history-of-earth-day/

Egan, Dan. 2017. *The Death and Life of the Great Lake. New York: W.W. Norton & Company*

Ellen MacArthur Foundation. n.d. "Circular Economy Introduction: Overview." Last Accessed 2023. https://www.ellenmacarthurfoundation.org/circular-economy/overview/concept

Encyclopedia Britannica. 1998. "Thomas Malthus." Economics & Economics Systems. Last Modified February 28, 2023. https://www.britannica.com/biography/Thomas-Malthus

Environmental and Energy Study Institute. 2017. "Biogas: Converting Waste to Energy, 2017 Fact Sheet." Last Accessed 2023. https://www.eesi.org/files/FactSheet_Biogas_2017.09.pdf

Environmental Research & Education Foundation. 2017. "Analysis of MSW Landfill Tipping Fees." Published April 1, 2017. https://erefdn.org/wp-content/uploads/2017/12/EREF-MSWLF-Tip-Fees-2017.pdf

Friedman, L. Thomas. 2016. *Thank You For Being Late: An Optimist's Guide to Thriving in the Age of Accelerations.* New York: Picador Paper; Illustrated Edition

Global Development Research Center. n.d. "Life Cycle Analysis and Assessment." Last Accessed 2023. https://www.gdrc.org/uem/lca/life-cycle.html

Global Footprint Network. 2018. "Ecological Footprint." How the Footprint Works. Last Accessed 2023. https://www.footprintnetwork.org/our-work/ecological-footprint/

Great Lakes Commission. n.d. "About the Lakes." Last Accessed 2023. https://www.glc.org/lakes/#:~:text=The%20lakes%20provide%20the%20backbone,billion%20annually%20for%20the%20region.

Green America. n.d. "Green Businesses Network Certification."Last Accessed 2023. https://www.greenamerica.org/green-america-green-business-certification

Green Dining Alliance. n.d. "The Mobile Mobius: A History of the Recycling Symbol." Last Accessed 2023. https://greendiningalliance.org/2016/03/the-mobile-mobius-a-history-of-the-recycling-symbol/

GreenSpec. n.d. "Copper Production and Environmental Impact." Last Accessed 2023. http://www.greenspec.co.uk/building-design/copper-production-environmental-impact/

GRS. 2018. "GRS - Global Recycle Standard." Last Accessed 2023. https://certifications.controlunion.com/en/certification-programs/certification-programs/grs-global-recycle-standard

Historic Fresno. n.d. "National Register of Historic Places: Fresno Sanitary Landfill (1937)." Last Accessed 2023. http://historicfresno.org/nrhp/landfill.htm

Habicht II, F. Henry. 1992. "Memorandum: EPA Definition of "Pollution Prevention."" EPA Office of the Administrator. Published May 28, 1992. https://www.epa.gov/sites/production/files/2015-07/documents/pollprev.pdf

Husain, Ali. 2021. "On This Day in 1871: The Great Midwest Wildfires of 1871." Last Accessed 2023. https://www.weatherbugsc.com/news/On-This-Day-in-1871-The-Great-Midwest-Wildfires-o

Huttner, Paul. 2015. "Water is the New Oil: Piping Lake Superior Water West." *Minnesota Public Radio*. April 22, 2015. https://blogs.mprnews.org/updraft/2015/04/water-is-the-new-oil-piping-lake-superior-water-west/

Investopedia. n.d. "Consumer Spending: Definition, Measurement and Importance." Last modified October 31, 2021. https://www.investopedia.com/terms/c/consumer-spending.asp

ISO. n.d. "Standards." Last Accessed 2023. https://www.iso.org/standards.html

Jefferson Recycling. 2017. "Exploring the History of Recycling – A Brief Timeline." Published April 13, 2017. https://www.jefferson-recycling.com/2017/04/exploring-history-recycling-brief-timeline/

Joseph S. Pete. 2017. Chicago Suing U.S. Steel Over Lake Michigan Discharge. *The Times of Northwest Indiana*. November 20, 2017. https://www.nwitimes.com/business/local/chicago-suing-u-s-steel-over-lake-michigan-discharge/article_16bb69bb-9cb5-5e2a-85fa-ff928894708f.html

Kent County Department of Public Works. n.d. "Recycling & Education Center." Last Accessed 2023. http://www.reiminetrash.org/facilities/recycling-education-center/

Lean Enterprise Institute, n.d. "Lean Thinking and Practice." Last Accessed 2023. https://www.lean.org/lexicon-terms/lean-thinking-and-practice/

Marion, Jeffrey. 2014. *LEAVE NO TRACE in the Outdoors*. Maryland: Stackpole Books

Market Business News. 2018. "What is Scarcity? Definition and Meaning." Financial Glossary. Last Accessed 2023. https://marketbusinessnews.com/financial-glossary/scarcity-definition-meaning/#:~:text=%E2%80%9CIn%20economic%20terms%2C%20it%20means,countries%20as%20in%20poor%20ones.%E2%80%9D

Merriam-Webster. n.d. "Definition: Gross Domestic Product." Last Accessed 2023. https://www.merriam-webster.com/dictionary/gross%20domestic%20product

Michigan Department of Agriculture and Rural Development. 2017. "Environmental Stewardship Division 2017 Annual Report: Maeap Water Use Reporting." Last Accessed 2023. https://www.michigan.gov/-/media/Project/Websites/mdard/documents/annual-reports/esd/2017_esd_annual_report.pdf?rev=24c575ffbc0145519b600c501a51885b

Michigan Department of Agriculture and Rural Development. 2017. "Agriculture Development Division Annual Report: Fiscal Year 2017." Last Accessed 2023. https://www.michigan.gov/ /media/Project/Websites/mdard/documents/annual-reports/agd/2017_agd_annual_report.pdf?rev=eefed0b16ed841688d4da3d6ce328131

Michigan Department of Environmental Quality. 2016. Measuring Recycling In The State Of Michigan: 2014 Recycling Rate. Published November 1, 2016. https://www.michigan.gov/egle/-/media/Project/Websites/egle/Documents/Programs/MMD/Recycling/Michigan-Recycling-Measurement-Report-November-2016.pdf?rev=12eb4b30b29c495cbae69cccea61ef81

Michigan Department of Environmental Quality. 2018. "Report of Solid Waste Landfilled in Michigan." Published January 31, 2018. https://cdm16110.contentdm.oclc.org/digital/collection/p9006coll4/id/188764/rec/22

Michigan Department of Natural Resources. 2018. "Accomplishments Report: Forest Resources Division Fiscal Year 2018." Last Accessed 2023. https://www.michigan.gov/-/media/Project/Websites/dnr/Documents/FRD/Mgt/FRD_Acc_Report.pdf?rev=c043385605eb4d57a2b6676810a0fe3a

Michigan Legislature. 2017. "Natural Resources And Environmental Protection Act (Excerpt) Act 451 of 1994." Last Accessed 2023. http://www.legislature.mi.gov/(S(fvhujoswppciadddwhco54me))/mileg.aspx?page=getObject&objectName=mcl-324-8905a

MSU Department of Geography and College of Social Science, n.d. "Hydrocarbons: Oil and Gas." Last Accessed 2023. http://geo.msu.edu/extra/geogmich/oil&gas.html

National Conference of State Legislatures. n.d. "States With Littering Penalties." Last Modified March 21, 2022. http://www.ncsl.org/research/environment-and-natural-resources/states-with-littering-penalties.aspx

National Geographic. 2018. "We Made Plastic. We Depend on it. Now Were Drowning in it." Published May 16, 2018 by Laura Parker. https://www.nationalgeographic.com/magazine/article/plastic-planet-waste-pollution-trash-crisis

National Geographic. 2016. "The Green, Brown, and Beautiful Story of Compost." Published September 9, 2016 by Aaron Sidder. https://www.nationalgeographic.com/culture/article/compost--a-history-in-green-and-brown

National Geographic. n.d. "Save the Plankton, Breathe Freely." Last Accessed 2023. https://www.nationalgeographic.org/activity/save-the-plankton-breathe-freely/

National Ocean Service by NOAA. n.d. "What are Microplastics?" Last Modified January 26, 2023. https://oceanservice.noaa.gov/facts/microplastics.html

National Ocean Service by NOAA. n.d. "Ocean Facts: How Much Water is in the ocean?" Last Modified January, 20, 2023. https://oceanservice.noaa.gov/facts/oceanwater.html

National Waste & Recycling Association. 2016. "Economic Data and Impact for the Waste & Recycling Industry." Last Accessed 2023. https://wasterecycling.org/wpcontent/uploads/2020/09/Economic_Data_Industry_2016.pdf

Northeast-Midwest State Foresters Alliance. n.d. "Michigan Forestry In the Great Lakes State." Last Accessed 2023. http://www.northeasternforests.org/content/michigan

Oceana. nd. "Restoring The Oceans Could Feed 1 Billion People A Healthy Seafood Meal Each Day." Last Accessed 2023. https://oceana.org/our-campaigns/feedtheworld/

Oxford Learner's Dictionary. n.d. "Sustain." Definition of Sustain Verb from the Oxford Advanced Learner's Dictionary. Last Accessed 2023. https://www.oxfordlearnersdictionaries.com/us/definition/english/sustain

Pianka, R. Eric. n.d. "Land." Last Accessed 2023. http://www.zo.utexas.edu/courses/thoc/land.html

Portney, E. Kent. 2015. *Sustainability: MIT Press Essential Knowledge Series.* Cambridge: The MIT Press.

Schultz, Wesley P. and Steven R. Stein 2009. "Executive Summary: Litter in America, 2009 National Litter Research Findings and Reccomendations." Keep America Beautiful, 2009 Report. https://keeplouisianabeautiful.org/wp-content/uploads/2015/09/Litter-in-America-Executive_Summary_-_FINAL.pdf

Science History Institute. n.d. "Science Matters: The Case of Plastics." Last Accessed 2023. https://www.sciencehistory.org/science-of-plastics

Seymour, S. Robert. 2016. "Managing an Aging Resource: Influence of Age on Leaf Area Index, Stemwood Growth, Growth Efficiency, and Carbon Sequestration of Eastern White Pine." Last Accessed 2023. https://nsrcforest.org/sites/default/files/uploads/seymour11full.pdf

Six Sigma. n.d. "About Six Sigma." Last Accessed 2023. https://www.6sigma.us/six-sigma.php

Teresa Hull and Linda Leask. 2000. "Dividing Alaska, 1867-2000: Changing Land Ownership and Management." Alaska Review of Social and Economic Conditions, Volume XXXII, no. 1 (Novemeber): Page 1. http://dnr.alaska.gov/commis/cacfa/documents/FOSDocuments/AKReviewOfSocialAndEconomicConditionsNov2000.pdf

Terracycle. n.d. "How Terracycle is Eliminating the Idea of Waste." Last Accessed 2023. https://www.terracycle.com/en-US/pages/how-terracycle

The Natural Step. n.d. "Approach: Accelerating Change." Last Accessed 2023. https://thenaturalstep.org/approach/

The Ocean Cleanup. nd. "The Great Pacific Garbage Patch." Last Accessed 2023. https://www.theoceancleanup.com/great-pacific-garbage-patch/

The Rewilding Institute. n.d. "Dave Foreman | Founder of The Rewilding Institute." Last Accessed 2023. https://rewilding.org/dave-foreman/

The World Bank. n.d. "Blue Economy: Oceans, Fisheries and Coastal Economies." Last Modified October 13, 2022. https://www.worldbank.org/en/topic/oceans-fisheries-and-coastal-economies

The World Energy Foundation. 2015. "A Brief History of Sustainability." Published March 24, 2015. https://medium.com/@twef/a-brief-history-of-sustainability-85a37d50d870

Townsend, Timothy G., and Jon Powell, Jain Pradeep, Debra Reinhart, Thabet Tolaymat, and Qiyong Xu. 2015. *Sustainable Practices for Landfill Design and Operation: Waste Management Principles and Practice.* New York: Springer Nature

TRUE. 2017. "TRUE Rating System." Modified October 1, 2022. https://true.gbci.org/sites/default/files/resources/TRUE-Rating-System-2022.pdf

TRUE. n.d. "TRUE Projects." Last Accessed 2023. https://true.gbci.org/projects

U.S. Department of Agriculture Natural Resources Conservation Service. n.d. "National Resources Inventory (NRI): A statistical survey of land use and natural resource conditions and trends on U.S. non-Federal lands." Last Accessed 2023. https://www.nrcs.usda.gov/nri

U.S. Department of Energy. n.d. "The National Environmental Policy Act of 1969, as Amended." NEPA of 1969. Last Amended September 13, 1982. https://www.energy.gov/nepa/downloads/national-environmental-policy-act-1969

U.S. Department of Justice. 2016. "National Sources of Law Enforcement Employement Data." Last Accessed 2023. https://bjs.ojp.gov/library/publications/national-sources-law-enforcement-employment-data

U.S. Energy Information Administration. 2018. "Biomass Explained, Waste-to-Energy (Municipal Solid Waste)." Last Modified October 31, 2022. https://www.eia.gov/energyexplained/?page=biomass_waste_to_energy

U.S. Environmental Protection Agency. 2018. "Advancing Sustainable Materials Management: Facts and Figures Report, 2018 Fact Sheet." Published Decemeber of 2020. https://www.epa.gov/facts-and-figures-about-materials-waste-and-recycling/advancing-sustainable-materials-management

U.S. Environmental Protection Agency. n.d. "Air Emissions from MSW Combustion Facilities." Last Modified March 29, 2016. https://archive.epa.gov/epawaste/nonhaz/municipal/web/html/airem.html

U.S. Environmental Protection Agency. n.d. "Criteria for the Definition of Solid Waste and Solid and Hazardous Waste Exclusions." Last Modified January 26, 2023. https://www.epa.gov/hw/criteria-definition-solid-waste-and-solid-and-hazardous-waste-exclusions#solidwaste

U.S. Environmental Protection Agency. n.d. "Energy Recovery from the Combustion of Municipal Solid Waste (MSW)." Last Modified February 9, 2023. https://www.epa.gov/smm/energy-recovery-combustion-municipal-solid-waste-msw

U.S. Environmental Protection Agency. n.d. "EPA History: Resource Conservation and Recovery Act." Last Modified June 27, 2022. https://www.epa.gov/history/epa-history-resource-conservation-and-recovery-act

U.S. Environmental Protection Agency. n.d. "Landfill Methane Outreach Program (LMOP): Landfill Technical Data." Last Modified September 6, 2022. https://www.epa.gov/lmop/landfill-technical-data

U.S. Environmental Protection Agency. n.d. "Municipal Solid Waste." Last Modified March 29, 2016. https://archive.epa.gov/epawaste/nonhaz/municipal/web/html/

U.S. Small Business Administration. 2018. "2018 Small Business Profile." United States. Last Accessed 2023. https://cdn.advocacy.sba.gov/wp-content/uploads/2018/11/23101710/2018-Small-Business-Profiles-US.pdf?utm_medium=email&utm_source=govdelivery

United Nations Department of Economic and Social Affairs. 2013. "World Population Projected to Reach 9.6 billion by 2050." Last Accessed 2023. https://www.un.org/en/development/desa/news/population/un-report-world-population-projected-to-reach-9-6-billion-by-2050.html

United Nations. n.d. "The Sustainable Development Agenda." Last Accessed 2023. https://www.un.org/sustainabledevelopment/development-agenda/

United Nations Development Programme. n.d. "What is Human Development." Last Accessed 2023. https://hdr.undp.org/about/human-development

West Michigan Sustainable Business Forum. 2016. "Economic Impact Potential and Characterization of MSW in Michigan, 2016 Report." Last Accessed 2023. https://wmsbf.files.wordpress.com/2016/04/michigan-msw-characterization-and-valuation-20161.pdf

World Resources Institute. 2022. "Ocean-Based Carbon Dioxide Removal: 6 Key Questions, Answered." Published November 15, 2022. https://www.wri.org/insights/ocean-based-carbon-dioxide-removal#:~:text=We%20know%20the%20ocean%20is,more%20carbon%20than%20the%20atmosphere.

Zero Waste America. n.d. "Landfills." Last Accessed 2023. http://www.zerowasteamerica.org/Landfills.htm

Zero Waste International Alliance. n.d. "Zero Waste Definition." Last Modified December 20, 2018. https://zwia.org/zero-waste-definition/

About the Author

Tyler Kanczuzewski is the vice president of marketing and sustainability, a board member, and an investor at Inovateus Solar. He also serves as the sustainability manager of Logistick, Inc. Originally from South Bend, Indiana, he immersed himself in the Grand Rapids and West Michigan sustainability community while working on his MBA (with a sustainable emphasis) at Grand Valley State University, where he graduated in 2019. He has led company efforts in stewardship and sustainable practices for both Inovateus Solar and Logistick since.

Printed in the United States
by Baker & Taylor Publisher Services